Flooring Components

Flooring Components

Building
Performance
Group

BPG Building Fabric Component Life Manual

First published 1999 by E & FN Spon,
11 New Fetter Lane, London EC4P 4EE

Simultaneously published in the USA and Canada
by E & FN Spon
29 West 35th Street, New York, NY 10001

E & FN Spon is an imprint of the Taylor and Francis Group

© Building Performance Group Limited, 1999.

Designed and typeset by AOK Group Company Limited, Orpington, Kent
Typeset in News Gothic

Printed and bound by George Overs Ltd, Rugby, Warwickshire

This manual was commissioned by the Defence Estates Organisation of the UK Ministry of Defence, and was compiled by durability specialists, Building Performance Group Limited. The principal author and editor was Gary Moss. Individual sections were prepared by Dianne McGlynn, Richard Newton and The Bonshor Partnership.

All technical enquiries should be directed to:

Building Performance Group Ltd.
Grosvenor House
141-143 Drury Lane
London WC2B 5TS
Tel: 0171 240 8070
Fax: 0171 836 4306
E-mail: enquiries@bpg-uk.com
Web Site: www.bpg-uk.com

The authors acknowledge the free license of the Defence Estates Organisation to reproduce copies of this Manual for UK government use, under contract WS13/3599.

A catalogue record for this book is available from the British Library.

ISBN 0-419-25510-9
ISSN 1468-5639

BPG

BPG	Contents of BPG Manual	Page

Contents

Foreword

Any building is a long-term high-cost asset which doesn't stop consuming money after the initial capital cost. In order to meet the demands of cost effective maintenance the design team, the maintenance team, the facilities manager and the clients need to know how components and facilities perform and decay over their lifetime.

Just like human beings, buildings require regular health checks, cleaning, proper protection, annual maintenance, periodic maintenance, refurbishment and, eventually, replacement! If we don't maintain our body, premature obsolescence and decay is the result. Buildings and components are the same, but have no way of communicating the problem until it is almost too late!

This manual will help both the designer, and the organisation responsible for maintaining a facility, to make effective choices. It sets out clearly the benchmarks for building component quality and the life expectancy together with the maintenance requirements. Such a document is badly needed by the industry to help plan both short term and long term maintenance and avoid nasty surprises. The work that we have undertaken here at Reading University in the area of whole life appraisal has desperately needed the type of information contained in this manual.

This publication supplements the HAPM Components Life Manual produced by BPG for Housing Association Property Mutual Ltd, a mutual insurance company providing latent defects cover for housing associations engaged in development. In common with the HAPM publication, the manual is clearly laid out and easy to follow. It is simple but profound. It is designed to be understood by everyone in the design and construction process, from the professional through to the tradesperson. It never leaves the reader wondering where to proceed next. Documents such as this are not easy to write, they take a great deal of time and professional skill and the Building Performance Group are to be congratulated on producing an excellent publication which is badly needed. Special thanks should go to the Defence Estates Organisation, for sponsoring the work.

I hope the work will not stop at this point but will progress forward, such that the industry builds up a very comprehensive manual of how project components can degrade and how we can forecast maintenance and keep our built facilities in good working order. The work shows both great foresight and a desire to move the topic forward.

The industry must take this document seriously it provides an important health plan and health check.

Professor Roger Flanagan

Department of Construction Management & Engineering
The University of Reading

Acknowledgements

The authors are indebted to the many organisations and individuals who assisted with the preparation of this manual, and without whom the task would have been considerably more difficult to complete. Particular thanks are due to the numerous manufacturers, suppliers, trade and research organisations who willingly contributed their time, data and support to this project.

BPG

Introduction

This Manual was prepared by durability specialists Building Performance Group Limited under the sponsorship of the Defence Estates Organisation of the Ministry of Defence. It arose from a recognised need for comparative data on the durability and maintenance of building components, materials and assemblies used in typical commercial, industrial and public building types. To date, such data has been largely unavailable in a published form.

The Manual was conceived as a companion volume to the existing HAPM Component Life Manual which schedules "insured lives" and maintenance requirements for in excess of 500 domestic building components, materials and assemblies. Published in 1992, the HAPM Manual was prepared to underpin the Housing Association Property Mutual (HAPM) 35 year latent defects insurance scheme for housing associations. HAPM insures components in respect of premature failure and it is therefore necessary to have "insured lives" for those components with a life expectancy of less than the 35 year insurance period. Building Performance Group subsidiary, Construction Audit Limited, was therefore commissioned to prepare schedules of the components used together with their assigned "insured lives". Since there was little alternative published information on the predicted lives of building components, it was decided to publish the HAPM schedules for the benefit of building owners, designers and specifiers.

For the purpose of consistency the format of the HAPM Manual, including the concept of "insured lives", has been adopted for this new publication. Consequently, the lives quoted are somewhat cautious and readers may sometimes wish to apply their own adjustment factors before using the data for other purposes such as maintenance scheduling and life cycle costing exercises.

In common with the HAPM Manual, components have been divided into a number of types and sub-types. For each sub-type there are a number of generic component descriptions, or 'quality benchmarks', each of which has an assigned life class. The life classes embrace good practice, a normal amount of maintenance and typical exposure conditions. Under this system users are able to assign the appropriate life class to a particular component specification or product.

It is important, however, to bear in mind that this Manual in no way seeks to limit the choices of building components available to designers and specifiers. It may be quite valid to select a short life component where access for maintenance or replacement will be straightforward and higher costs in use are acceptable. The need for replacement may be dictated by factors unconnected with the failure of a component, such as non-availability of spare parts or a desire to improve standards. Nevertheless, the ability to differentiate between qualities of component in order to assign a longer life to those which can be identified as having enhanced durability characteristics is considered to be a powerful tool for specifiers. Component descriptions have therefore been ranked in order of durability, which could also be termed "quality" in a particular and a limited sense.

The list of components covered by this Manual has been selected to include those which are commonly used in commercial, industrial and public building types and which are not covered by the existing HAPM Manual for domestic components. It is envisaged that twice yearly supplements will be issued on a subscription basis to update the existing data and to add sections on additional building components which have not been included in the current edition. A further manual is under preparation (also sponsored by the Defence Estates Organisation), which will provide durability data for a range of mechanical and electrical services components.

Origins of the durability data

The data in this manual has been assembled by Building Performance Group using a rigorous research process developed from the company's 10 years' experience in this field. The durability assessment process was originally developed to provide insured life assessments for the HAPM latent defects insurance scheme on its inception in 1990. It has since been refined through a four year collaborative research programme with the University of the West of England and through research and consultancy work for clients such as the National Housing Federation, Building Research Establishment, Defence Estates Organisation and British Standards Institute.

For each building component included in the manual, durability data has been assembled from a diverse range of sources including, where relevant:

➤ Manufacturers and Suppliers
➤ Manufacturer and Trade Associations
➤ Research Organisations such as BRE, CIRIA, TRADA
➤ Test Houses and Certification Bodies
➤ British and European Standards
➤ Published research and conference papers
➤ Technical press
➤ Existing published sources of component life data

Wherever possible the above data has been supplemented by feedback from the performance of buildings in use and on the most commonly encountered causes of deterioration and failure. Analysis of this data has enabled a series of 'quality benchmarks', or levels of performance, to be identified for each building component under consideration. Life assessments and statements of minimum required maintenance have been assigned to each benchmark.

The prediction of building component lives is not an exact science - it is acknowledged that the data presented in this manual can be no more than an informed estimate based on the information available at the time of writing. The authors keep under regular review changes in British and European Standards, and developments by testing and research organisations and product manufacturers. The component descriptions and life assessments will be reviewed periodically to take account of such developments and any necessary changes will be incorporated into the six monthly updating service. The authors would also welcome any comments or feedback from users of the manual.

About the authors

Building Performance Group is a multi-disciplinary consultancy with an international reputation for its work in the field of building component durability. Its publications on the subject include the well-known HAPM Component Life Manual (E & FN Spon, 1992), a Ministry of Defence guide to durability auditing (BPG, 1999), and a BRE published defects avoidance manual (BRE, 1991). The company is represented on a number of industry research and standards committees and has drafted parts 1 and 3 of a forthcoming international standard on service life planning for buildings.

As technical auditor to the Housing Association Property Mutual latent defects insurance scheme, BPG (through its subsidiary Construction Audit Limited) has unique access to data on the performance in use of more than 100,000 housing units. This data is supplemented by the company's extensive experience in the fields of condition surveys and inspections, defects investigations, maintenance planning and refurbishment of a broad range of building types.

Use of the Manual

Structure

The Manual is divided into a number of Sections, each representing a major building element.

Section One: Flooring Components
Section Two: Walling and Cladding Components
Section Three: Roofing Components
Section Four: Doors, Windows, and Joinery Components
Section Five: Stairs and Balustrades
Section Six: Internal Fixtures and Fittings

Component Types and Sub-types

Component types such as rooflights, doors and floor screeds are assigned to the appropriate Section and are separated into component sub-types which are generally differentiated by material. Within each sub-type, components are described by different qualities (or 'benchmarks') of specification and each specification is assigned a life assessment ranging from 5 to 35+ years. There is also a life class designated "U" (meaning unclassified) which is used where the component does not comply with relevant British Standards or is unsuitable for the purpose specified, or where there is insufficient information presented to allocate a life class.

An introduction page is provided for each component type in the manual, outlining the scope of coverage for the component and providing a bibliography of all British and European Standards and other references cited.

The Life Classes

To enable this Manual to be used alongside the HAPM Component Life Manual the component life assessments quoted are analogous to "insurance lives", which represent somewhat cautious assessments of durability. Since the maximum insured life under the HAPM latent defects insurance scheme is 35 years, an additional classification of 35+ years has been adopted. This classification is assigned to components which are expected to last beyond 35 years and which can often be expected to last for the typical 60 year design life of a building.

Location

Where the service life of a component is likely to vary depending on its location in the building, (e.g. internal, external), alternative life assessments are given for different locations.

Where the life of a component is unaffected by its location within the building its location is described as "general".

Adjustment Factors

The service life of a component can often be affected by local conditions such as marine environments, polluted/ industrial atmosphere and frost pockets. Positive and negative adjustment factors are provided to take account of such local conditions. Definitions for environmental conditions which are used throughout the Manual have been included as an appendix for guidance.

Assumptions

In assigning life assessments to components certain general assumptions have been made throughout the manual in addition to specific assumptions which relate to particular components.

General assumptions which apply throughout include installation in accordance with manufacturers' directions, good practice, relevant Codes of Practice and British/European Standards and the use of appropriate design details. In addition, where a third party assurance certification scheme applies, such as a British Board of Agreement (BBA) certificate, installation is assumed to comply with the conditions of the relevant certificate.

Specific assumptions, highlighting particular facets of detailing or available guidance on installation, have been made in estimating the durability of specific components. These may reflect, for example, compliance with a trade standard for installation, or ensuring that incompatible materials are adequately separated. Such assumptions are stated on the relevant pages in the Manual.

Non-compliance with these assumptions may have a negative effect on component durability.

Use of the Manual

Maintenance

Life assessments are assigned to components on the assumption that a certain minimum level of maintenance will be carried out. Each component data sheet specifies the minimum required maintenance. Any manufacturer's specific maintenance requirements should also be carried out.

Failure to adequately maintain certain components e.g. roof coverings may result in premature failure of other components which they should protect. It is assumed that small items of normal maintenance will be carried out in time to ensure that consequential damage to other components does not result.

Notes

Where appropriate, additional notes are provided on matters such as further sources of information, incompatibilities, unsuitable applications or materials.

Key Failure Modes and Key Durability Issues

Lists of key failure modes and key durability issues are provided for each component, to explain the reasoning behind the life assessments and to highlight the key determinants of durability in practice.

Index

An alphabetical index has been provided of all components included in the Manual.

Appendices

The following information is included in appendices:

- A list of all abbreviations used throughout the Manual.
- Definitions of environmental exposure conditions cited in the adjustment factors.
- List of domestic building components included in the HAPM Component Life Manual.

List of components

1 – Flooring components

2 – Walling & cladding components

3 – Roofing components

List of components

BPG List of components in section order

Floor screeds & toppings

Scope

This section provides data on floor screeds and toppings commonly used in commercial and industrial building types. For the purposes of this document, a screed is defined as a layer of cement:sand or other material laid on the structural floor to receive another finish such as carpet or tiles. A topping is defined as a layer of high strength concrete which provides the wearing surface to a concrete base.

The following component sub-types are included within this section:

	Page
• In-situ terrazzo toppings	1.1
• High strength concrete toppings	1.1a
• Direct finished concrete slabs	1.1a
• Proprietary cementitious toppings	1.1b
• Proprietary resin-based toppings	1.1c

This section does not include data on precast flooring units (eg terrazzo tiles/slabs), proprietary resin-based coatings of applied thickness less than 0.5mm, or bituminous toppings. Magnesium oxychloride toppings, which are no longer in common use, are also excluded. Domestic floor screeds, including cementitious, calcium sulfate and mastic asphalt, are covered in Section 1 of the HAPM Component Life Manual.

It should be noted that there is an extensive range of proprietary floor screeds and toppings currently available to specifiers, and that an attempt has been made to categorise the principal types only. Since many systems are designed for highly specific applications, it is essential to seek the advice of the specialist manufacturer or supplier prior to making a final selection.

Standards cited

BS 12: 1996	Specification for Portland Cement
BS 63	Road aggregates
Part 1: 1987(1994)	Specification for single-sized aggregate for general purposes
BS 146: 1996	Specification for Portland blast furnace cements
BS 882: 1992	Specification for aggregates from natural sources for concrete
BS 915	Specification for high alumina cement
Part 2: 1972 (1983)	Metric Units
BS 1014:1975 (1992)	Specification for pigments for Portland cement and Portland cement products
BS 1047: 1983	Specification for air-cooled blastfurnace slag aggregate for use in construction
BS 1199: 1976 (1996)	Specifications for building sands from natural sources
BS 1521: 1972 (1994)	Specification for waterproofing building papers
BS 4027: 1996	Specification for sulfate-resisting Portland cement
BS 5075	Concrete admixtures
Part 1: 1982	Specification for accelerating admixtures, retarding admixtures and water reducing admixtures
Part 2: 1982	Specification for air-entraining admixtures
Part 3: 1985	Specification for super-plasticising admixtures
BS 6588: 1996	Specification for Portland pulverised-fuel ash cements
BS 8110:	Structural use of concrete
Part 1: 1997	Code of Practice for design and construction
BS 8204	Screeds, bases and in-situ floorings
Part 1: 1987	Code of Practice for concrete bases and screeds to receive in-situ floorings
Part 2: 1987	Code of Practice for concrete wearing surfaces
Part 3: 1993	Code of Practice for polymer modified cementitious wearing surfaces
Part 4: 1993	Code of Practice for terrazzo wearing surfaces

Other references/information sources

National Federation of Terrazzo-Marble & Mosaic Specialists.

HAPM Component Life Manual pages 1.17 - 1.17b (domestic floor screeds).

Federation of Specialist Contractors & Material Suppliers to the Construction Industry (FeRFA) Application Guides:

No.4: Synthetic Resin Flooring

No.6: Polymer Flooring Guide

LOCATIONS - Ground/Intermediate floors

1 - Flooring Components

General | **Description**

In-situ terrazzo toppings

35+
In-situ terrazzo topping to BS 8204:Part 4. Crushed marble aggregate minimum size 3mm, maximum 25mm. Cement:aggregate mix to be no greater than 1:2 by volume, except for thin sections where mix may be increased to 2:3. Laid on bonded screed.

30
In-situ terrazzo topping to BS 8204:Part 4. Crushed marble aggregate minimum size 3mm, maximum 25mm. Cement:aggregate mix to be no greater than 1:2 by volume, except for thin sections where mix may be increased to 2:3. Laid on unbonded screed/direct on concrete base.

U1
Unclassified, ie in-situ terrazzo topping not to BS 8204:Part 4.

Maintenance

Regular washing with warm water and a suitable neutral sulphate-free detergent. Avoid regular use of organic solvent or alkaline (pH > 9) detergent. Minor repairs should be carried out promptly.

Regular washing with warm water and a suitable neutral sulphate-free detergent. Avoid regular use of organic solvent or alkaline (pH > 9) detergent. Minor repairs should be carried out promptly.

Unclassified

For Adjustment factors, Assumptions, Notes, Key failure modes and Key durability issues please see overleaf.

1.1

Floor screeds & toppings

LOCATIONS - Ground/Intermediate floors

Adjustment Factors

None.

Assumptions

Minimum topping thickness 15mm for aggregates of 10mm or less. Increase thickness proportionately for larger aggregates. Skirtings should be minimum 6mm thick terrazzo, laid on minimum 10mm screed. Minimum terrazzo thickness for stairs: 15mm (treads), 10mm (risers), 6mm (strings).

Movement joints in the concrete base must be carried through the screed and terrazzo flooring to the floor surface. Proprietary metal edged movement joints should also be used around columns, between terrazzo and other floor coverings, centrally over supporting beams and walls of suspended structural floors.

Terrazzo flooring should be divided into panels generally not exceeding 1m² in area, separated by plastic or metal dividing strips. Length of panel should be no greater than twice the width. Dividing strips to be brass, aluminium alloy or plastics.

Any existing floor base must be sound and free from cracking or hollowness. Concrete screeds laid to receive terrazzo finish to be prepared in accordance with BS 8204:Parts 1 and 4. The base for the terrazzo should be provided with a float finish with a surface regularity of class SR3 as given in Table 2 of BS 8204: Part 1. Screed mix, thickness etc as specified in HAPM Component Life Manual, P.1.17.

Topping is not designed to resist hydrostatic pressure, therefore an independent damp-proof membrane may be required in ground floors.

Cement to be grey or white Portland Cement to BS 12 class 42.5N.

Sand to be to BS 882 or BS 1199.

Marble aggregates to be free from fines and dust, and angular as distinct from elongated and flaky. Mechanical properties of terrazzo aggregates to be within the ranges given in Table 1 of BS 8204:Part 4.

Aggregates should not contain any deleterious materials (eg coal, lignite and iron pirites) in sufficient quantity to adversely affect the surface finish.

Pigments/colouring agents to conform to BS 1014.

The following admixtures may be used: air-entraining (to BS 5075:Part 2), water reducing (BS 5075:Part 1), super-plasticising (BS 5075:Part 3), accelerating (BS 5075:Part 1), retarding (BS 5075:Part 1). NB: Admixtures containing calcium chloride should not be used in the vicinity of reinforced concrete or screeds containing embedded metal.

Additional materials such as polymer dispersions, microsillica fume, pulverised fuel ash may be used in the screed/topping by agreement with the specialist installer.

Notes

Whilst it is preferable for terrazzo toppings to be laid on a bonded screed, this will not always be possible. In situations where structural movement is expected (eg from settlement, expansion/ contraction or vibration), or where terrazzo is laid over a precast concrete floor, the screed should be laid on an isolating membrane of minimum 500 gauge plastic sheeting or BS 1521 building paper. Unbonded screeds should be a minimum of 60mm thick, and with a light steel fabric reinforcement placed at mid-thickness.

Regular use of scrub and rinse cleaning machines fitted with abrasive pads, or machines with hard plastic bristles, may cause damage to the surface. The frequent use of highly acidic/alkaline cleaning agents may also lead to surface damage. Some disinfectants can cause permanent discolouration.

Further information on the installation and maintenance of terrazzo flooring can be obtained from the National Federation of Terrazzo-Marble & Mosaic Specialists.

Key failure modes

Surface crazing, cracking, curling/lifting, loss of cement binder.

Key durability issues

Frequency/location of movement joints/divider strips, topping thickness, substrate preparation.

Floor screeds & _____ toppings

BPG | 1 - Flooring Components

General	Description	Maintenance
High strength concrete toppings (including granolithic)		
35	High strength topping to BS 8204:Part 2. Maximum 10mm single size coarse aggregate in accordance with table 4 of BS 882. Topping laid monolithically with base. Minimum topping thickness 15mm +/- 5mm.	Regular washing with water. Grease stains etc should be removed using aqueous solutions of alkaline salts (eg caustic soda, sodium metasilicate), or proprietary detergent compositions.
25	High strength topping to BS 8204:Part 2. Maximum 10mm single size coarse aggregate in accordance with table 4 of BS 882. Topping bonded to a set and hardened base. Minimum topping thickness 20mm +/- 5mm.	Regular washing with water. Grease stains etc should be removed using aqueous solutions of alkaline salts (eg caustic soda, sodium metasilicate), or proprietary detergent compositions.
20	High strength topping to BS 8204:Part 2. Maximum 10mm single size coarse aggregate in accordance with table 4 of BS 882. Laid as a separate unbonded overslab. Minimum topping thickness 100mm +/- 5mm.	Regular washing with water. Grease stains etc should be removed using aqueous solutions of alkaline salts (eg caustic soda, sodium metasilicate), or proprietary detergent compositions.
U1	Unclassified, ie high strength topping not to BS 8204:Part 2, or thickness not appropriate to substrate.	Unclassified.
Direct finished concrete slabs		
35+	Direct finished concrete slab to BS 8204:Part 2. High alumina cement to BS 915 (not suitable for use in structural concrete or in monolithic toppings or Portland cement concrete).	Regular washing with water. Grease stains etc should be removed using aqueous solutions of alkaline salts (eg caustic soda, sodium metasilicate), or proprietary detergent compositions.
35+	Direct finished concrete slab to BS 8204:Part 2. Portland-blastfurnace cement to BS 146.	Regular washing with water. Grease stains etc should be removed using aqueous solutions of alkaline salts (eg caustic soda, sodium metasilicate), or proprietary detergent compositions.
35+	Direct finished concrete slab to BS 8204:Part 2. Portland pulverized-fuel ash cement to BS 6588.	Regular washing with water. Grease stains etc should be removed using aqueous solutions of alkaline salts (eg caustic soda, sodium metasilicate), or proprietary detergent compositions.
35+	Direct finished concrete slab to BS 8204:Part 2. Combination of Portland cement with granulated blastfurnace slag or pulverised fuel ash.	Regular washing with water. Grease stains etc should be removed using aqueous solutions of alkaline salts (eg caustic soda, sodium metasilicate), or proprietary detergent compositions.
35	Direct finished concrete slab to BS 8204:Part 2. Portland cement to BS 12.	Regular washing with water. Grease stains etc should be removed using aqueous solutions of alkaline salts (eg caustic soda, sodium metasilicate), or proprietary detergent compositions.
35	Direct finished concrete slab to BS 8204:Part 2. Sulphate-resisting Portland cement to BS 4027.	Regular washing with water. Grease stains etc should be removed using aqueous solutions of alkaline salts (eg caustic soda, sodium metasilicate), or proprietary detergent compositions.
U1	Unclassified, ie direct finished concrete slab not to BS 8204:Part 2.	Unclassified.

1.1a

For Adjustment factors, Assumptions, Notes, Key failure modes and Key durability issues please see overleaf.

Floor screeds & toppings

Adjustment Factors

Application of proprietary sprinkled finish of hard natural aggregates, metallic and synthetic aggregates to the plastic concrete surface: +5 years (direct finished slab only).

Flooring subject to severe abrasive conditions, eg traffic by loaded vehicles with steel or rigid plastics wheels: -10 years.

Prolonged exposure to acids, vegetable oils, fats or sugar solutions: -5 years.

Assumptions

Concrete substrate to be prepared in accordance with BS 8204:Part 1. Any existing floor base must be sound and free from cracking or hollowness.

Minimum topping thickness: 15mm +/- 5mm for monolithic construction, 20mm for separate bonded construction, 100mm for separate unbonded construction.

Minimum thickness of direct finished slabs: 100mm.

Design, including bay sizes and provision of movement joints, to be in accordance with BS 8204:Part 2.

Toppings are not designed to resist hydrostatic pressure, therefore an independent damp-proof membrane may be required in ground floors.

With bonded toppings, any movement joints in the concrete base slab must be carried through the topping to avoid cracking.

Cement to be one of the following: Portland cement to BS 12, Portland-blastfurnace cement to BS 146:Part 2, High alumina cement to BS 915, sulphate-resisting Portland cement to BS 4027, Portland pulverized-fuel ash cement to BS 6588, or a combination of Portland cement with granulated blastfurnace slag or pulverised fuel ash. High alumina cement should not be used in structural concrete or in monolithic toppings on Portland cement concrete.

Aggregates for direct finished surfaces should be selected from one of the following: aggregate complying with BS 882, coarse aggregates of blastfurnace slag complying with BS 1047, or coarse and fine aggregates of other (non-BS) types provided they are appropriate to the strength, density, shrinkage and durability of the concrete. 'All-in' aggregates should not be used.

Coarse aggregates for high strength concrete toppings should comply with BS 882. Fine aggregates for high strength concrete toppings should be natural sand complying with the grading limit C or M of Table 5 of BS 882.

Aggregates should not contain any deleterious materials in sufficient quantity to adversely affect the concrete surface (eg lignite and iron pirites).

Pigments/colouring agents to conform to BS 1014.

Admixtures to comply with the appropriate Part of BS 5075.

Ground supported base slabs to be laid on a sub-base of fully compacted granular material of maximum size 75mm, lightly blinded with sand, or with a lean mix concrete to form a flat surface.

Where the concrete is at risk from attack by aggressive soil water, either a damp-proof membrane should be provided under the slab, or an appropriate concrete grade used in accordance with BS 8110:Part 1.

Use of proprietary surface hardeners, sealers, or special aggregate/sprinkle finishes in accordance with suppliers' recommendations.

Pipes or conduits must not be buried in a topping.

Concrete for toppings on suspended in-situ concrete/ precast concrete floors which are designed to contribute to the structural strength of the floor should be designed in accordance with BS 8110.

Notes

BS 8204:Part 2 specifies four classes of abrasion resistance for concrete flooring, along with appropriate concrete grades, minimum cement content, aggregate types and finishing processes for each.

BS 8204:Part 2 high strength toppings include those commonly known as granolithic toppings.

The wearing qualities of direct finished slabs can be upgraded by the incorporation of a sprinkled finish of special aggregates and cement in the slab surface during the flat finishing process.

Power float finishes may not achieve an adequate finish (particularly in corners) if the area is to be subsequently carpeted.

Key failure modes

Surface crazing, cracking, curling/lifting, abrasion, loss of cement binder, chemical degradation (acids, fats).

Key durability issues

Curing period, abrasion resistance, frequency/location of movement joints, topping/screed thickness, substrate preparation.

Floor screeds & toppings

BPG	1 - Flooring Components	
General	**Description**	**Maintenance**
	Proprietary cementitious toppings	
25	BBA or other 3rd party certified polymer modified topping. Thickness and installation of topping in strict accordance with manufacturer's recommendations.	Regular washing with water. Removal of grease stains with aqueous solutions of alkaline salts (eg caustic soda, sodium metasilicate), or by using proprietary detergents. Minor repairs should be carried out promptly.
25	Preblended polymer-modified topping. Thickness and installation of topping in strict accordance with manufacturer's recommendations.	Regular washing with water. Removal of grease stains with aqueous solutions of alkaline salts (eg caustic soda, sodium metasilicate), or by using proprietary detergents. Minor repairs should be carried out promptly.
25	Proprietary metallic non-ferrous aggregate topping. Laid monolithically with concrete base/screed. Cement to comply with the requirements of BS 12, Class 42.5N, BS 915:Part 2, or BS 4027 Class 42.5N. Sand to BS 882, grades C or M, but with not more than 10% passing sieve size 75 micrometres. Mixing/laying in strict accordance with manufacturer's instructions.	Regular washing with water. Removal of grease stains with aqueous solutions of alkaline salts (eg caustic soda, sodium metasilicate), or by using proprietary detergents. Minor repairs should be carried out promptly.
20	Proprietary polymer-modified topping. Cement to comply with the requirements of BS 12, Class 42.5N, BS 915:Part 2, or BS 4027 Class 42.5N. Sand to BS 882, grades C or M, but with not more than 10% passing sieve size 75 micrometres. Coarse aggregate to table 2 of BS 63:Part 1 or table 3 of BS 882. Thickness and mix proportions in accordance with table 1 of BS 8204:Part 3.	Regular washing with water. Removal of grease stains with aqueous solutions of alkaline salts (eg caustic soda, sodium metasilicate), or by using proprietary detergents. Minor repairs should be carried out promptly.
20	Proprietary metallic non-ferrous aggregate topping. Laid monolithically with a minimum 25mm thick 1:1.5:2.5 concrete bonding layer. Cement to comply with the requirements of BS 12, Class 42.5N, BS 915:Part 2, or BS 4027 Class 42.5N. Sand to BS 882, grades C or M, but with not more than 10% passing sieve size 75 micrometres. Mixing/laying in strict accordance with manufacturer's instructions.	Regular washing with water. Removal of grease stains with aqueous solutions of alkaline salts (eg caustic soda, sodium metasilicate), or by using proprietary detergents. Minor repairs should be carried out promptly.
20	Proprietary epoxy reinforced cementitious topping. Minimum thickness 10mm. Mixing and application in strict accordance with manufacturer's instructions.	Regular washing with water. Removal of grease stains with aqueous solutions of alkaline salts (eg caustic soda, sodium metasilicate), or by using proprietary detergents. Minor repairs should be carried out promptly.
10	Proprietary epoxy reinforced cementitious topping. Minimum thickness 5mm. Mixing and application in strict accordance with manufacturer's instructions	Regular washing with water. Removal of grease stains with aqueous solutions of alkaline salts (eg caustic soda, sodium metasilicate), or by using proprietary detergents. Minor repairs should be carried out promptly.
U1	Unclassified, ie proprietary polymer-modified topping, cement not to BS 12/BS 915/BS 4027, sand not to BS 882, aggregate not to BS 63/BS 882 and/or thickness/mix proportions not to BS 8204:Part 3.	Unclassified.
U2	Unclassified, ie proprietary metallic aggregate topping not laid monolithically with concrete base or bonding layer, and/or cement not to BS 12/BS 915/BS 4027, sand not to BS 882, aggregate not to BS 63/BS 882, and/or not mixed/laid in accordance with manufacturer's instructions.	Unclassified.
U3	Unclassified, ie proprietary epoxy reinforced cementitious topping, thickness less than 1.5mm, and/or mixing/application not in accordance with manufacturer's instructions.	Unclassified.

1.1b	**For Adjustment factors, Assumptions, Notes, Key failure modes and Key durability issues please see overleaf.**

Floor screeds & toppings

LOCATIONS - Ground/Intermediate floors

1 - Flooring Components

Adjustment Factors

Prolonged exposure to organic solvents/wetting: -10 years (polymer modified only).

Flooring subject to severe abrasive conditions, eg traffic by loaded vehicles with steel or rigid plastics wheels: -10 years.

Assumptions

Design and installation in accordance with BS 8204:Part 3 (polymer modified only), and with manufacturer's recommendations.

Concrete base/screed to be prepared in accordance with BS 8204:Part 1.

Cement to be one of the following: Portland cement to BS 12, Portland-blastfurnace cement to BS 146, High alumina cement to BS 915, sulphate-resisting Portland cement to BS 4027, Portland pulverized-fuel ash cement to BS 6588, or a combination of Portland cement with granulated blastfurnace slag or pulverised fuel ash. High alumina cement should not be used in structural concrete or in monolithic toppings on Portland cement concrete.

Bonding agent to be applied prior to laying of polymer modified topping. Bonding agent to be in accordance with table 3 of BS 8204:Part 3.

All movement joints in the concrete base must be carried on through the topping. In addition, joints should be provided over supports on suspended floors, around the perimeters of the slab and around columns, manholes and fixed bases.

Toppings are not designed to resist hydrostatic pressure, therefore an independent damp-proof membrane may be required in ground floors.

Use of proprietary surface hardener/sealer where recommended by manufacturer Pigments to conform to BS 1014.

Concrete admixtures to conform to BS 5057:Part 1. Use of admixtures to be approved by polymer manufacturer/supplier.

Pipes or conduits must not be buried in a topping. Installer to be approved/nominated by manufacturer.

Notes

There is currently an extensive range of proprietary flooring systems available to specifiers, many of which are designed for highly specialised applications. It is also important to note that the performance of a given flooring system can vary considerably according to the chosen aggregate type, mix proportions and application thickness. It is therefore essential to seek the advice of the specialist manufacturer or supplier before making a final selection.

BS 8204:Part 3 defines three use classes of polymer modified flooring (light, medium and heavy duty), and provides recommendations for the thickness and mix proportions for each.

FeRFA application guide No.6 provides further guidance on the use of polymer-based flooring systems.

Key failure modes

Surface crazing, cracking, curling/lifting, abrasion, loss of cement binder, chemical degradation (acids, fats, organic solvents).

Key durability issues

Curing period, abrasion resistance, frequency/location of movement joints, topping/screed thickness, substrate preparation, selection of appropriate proprietary system.

Floor screeds & toppings_

LOCATIONS - Ground/Intermediate floors

1 - Flooring Components

General	Description	Maintenance
	Proprietary resin-based toppings	
15	Proprietary heavy duty topping comprising resin binder (epoxy/polyester/polyurethane) and graded natural/synthetic aggregates. Applied thickness in excess of 4mm. Mix proportions, application thickness, provision of surface sealing coat strictly in accordance with manufacturer's instructions.	Regular washing with water and neutral cleaning agents. Use of proprietary cleaning materials as recommended by flooring manufacturer. Avoid the use of acid or solvent based cleaning agents. Minor repairs should be carried out promptly.
10	Proprietary thin layer (self smoothing) resin based topping (epoxy/polyester/polyurethane). Minimum applied thickness 2mm. Mix proportions, application thickness, provision of surface sealing coat strictly in accordance with manufacturer's instructions.	Regular washing with water and neutral cleaning agents. Use of proprietary cleaning materials as recommended by flooring manufacturer. Avoid the use of acid or solvent based cleaning agents. Minor repairs should be carried out promptly.
5	Proprietary thin layer (self smoothing) resin based topping (epoxy/polyester/polyurethane). Minimum applied thickness 0.5mm. Mix proportions, application thickness, provision of surface sealing coat strictly in accordance with manufacturer's instructions.	Regular washing with water and neutral cleaning agents. Use of proprietary cleaning materials as recommended by flooring manufacturer. Avoid the use of acid or solvent based cleaning agents. Minor repairs should be carried out promptly.
U1	Proprietary resin based topping. Minimum applied thickness less than 0.5mm, and/or mix proportions/application thickness/surface sealing coat not in accordance with manufacturer's instructions.	Unclassified.

For Adjustment factors, Assumptions, Notes, Key failure modes and Key durability issues please see overleaf.

Floor screeds & toppings_

Adjustment Factors

Prolonged exposure to organic solvents: -10 years.

Use in loading bays/regular trafficking by vehicles/ trolleys using steel or nylon wheels: -5 years.

Prolonged exposure to UV light (direct sunlight/radiating through plastic rooflights): -5 years.

Assumptions

Selection of flooring system in accordance with manufacturer's recommendations.

Installation by specialist contractor.

Preparation of substrate (including application of primers/base coats) in accordance with BS 8204:Part 1 (concrete) and with the manufacturer's instructions.

All movement or other structural joints in the sub-floor/base must be continued up through the thickness of the topping.

Toppings are not designed to resist hydrostatic pressure, therefore an independent damp-proof membrane may be required in ground floors.

Notes

There is currently an extensive range of proprietary flooring systems available to specifiers, many of which are designed for highly specialised applications. It is also important to note that the performance of a given flooring system can vary considerably according to the chosen aggregate type, mix proportions and application thickness. It is therefore essential to seek the advice of the specialist manufacturer or supplier before making a final selection.

All resin-based products have a limited shelf-life, both before and after mixing. It is essential that the manufacturer's advice is adhered to.

Graded quartz sand (typically 0.1-0.3mm) may be added to a thin layer resin mix to produce a heavy duty coating. The required thickness will dictate the mix proportions and aggregate grain size. The manufacturer's guidance should always be sought.

The long term performance of resin-based toppings is often dependent upon the provision of a thin protective sealing coat. It is essential that any such coatings are maintained in good order.

Proprietary resin based coatings (ie applied thickness less than 0.5mm) are not included in this section..

FeRFA application guide No.4 provides further guidance on the use of resin-based flooring systems.

Key failure modes

Surface crazing, cracking, curling/lifting, abrasion, loss of cement binder, chemical degradation (acids, organic solvents. UV light).

Key durability issues

Abrasion resistance, provision of sealing coat, frequency/location of movement joints, topping thickness, substrate preparation, selection of appropriate proprietary system.

Raised access floors

Scope

This section provides data on raised access floors commonly used in office buildings to facilitate the free routing of cables and other mechanical and electrical services.

Raised access floors are generally defined as those laid over a void of less than 100mm (typically 74mm) and suitable for light cable applications. However, this section also includes platform floors, which are laid over a void of 100mm or more with a typical maximum of 600mm, and which may also be used as a plenum for air conditioning. These should not be confused with the mezzanine structures (also known as platform floors), which are commonly used for storage in factories and warehouses, and which are not covered in this section.

The following component sub-types are included within this section:

	Page
Metal pedestals	1.2
Other pedestal/support systems	1.2a
Floor panels	1.2b

Standards cited

BS 1088: 1966 (1988)	Specifications for plywood for marine craft
BS 1449 Part 1: 1991	Steel plate, sheet and strip
	Carbon and carbon-manganese plate, sheet and strip
BS 1490: 1988	Specification for aluminium and aluminium alloy ingots and castings for general engineering purposes
BS 1706: 1990 (1996)	Method for specifying electroplated coatings of zinc and cadmium on iron and steel
BS 6566 Plywood Part 8: 1985 (1991)	Specification for bond performance of veneer plywood
BS EN 312: 1997	Particle boards - Specifications
BS EN 485: 1994	Aluminium and aluminium alloys. Sheet, strip and plate
BS EN 10142: 1991	Specification for continuously hot-dip zinc coated low carbon steel sheet and strip for cold forming: technical delivery conditions
BS EN 10143: 1983	Continuously hot-dip metal coated steel sheet and strip. Tolerances on dimensions and shape

Other references/information sources

PSA MOB PF22 PS/SPU: 1992:	Platform Floors (Raised Access Floors) Performance Specification.
	Access Flooring Association.

TYPE

Raised access & _____
platform floors

LOCATIONS - Ground/Intermediate floors

SUB TYPES
Metal pedestals

1 - Flooring Components

General	Description	Maintenance
	Metal pedestals	
35+	Steel and/or aluminium composite adjustable pedestal forming part of system fully complying with the PSA MOB Specification. Steel galvanised to BS EN 10143 and passivated after fabrication. Aluminium to BS 1490. Comprising base plate of minimum size 10,000 mm^2 welded or rivetted to tube or box section column, threaded adjustable jack with locking nut and head including provision for positive location of panels. All pedestals fixed to sub-floor with manufacturer's recommended epoxy adhesive.	To be in accordance with manufacturers stated requirements. Manufacturers lifting device to be used when removing panels.
35	Steel and/or aluminium composite adjustable pedestal forming part of system fully complying with the PSA MOB Specification. Steel to BS 1449 and zinc plated to BS 1706. Aluminium to BS 1490. Comprising base plate of minimum size 10,000 mm^2 , threaded adjustable section including provision for positive location of panels. All pedestals fixed to sub-floor with manufacturer's recommended epoxy adhesive.	To be in accordance with manufacturers stated requirements. Manufacturers lifting device to be used when removing panels.
25	Steel and/or plastic composite adjustable pedestal forming part of system fully complying with the PSA Specification. Steel to BS 1449. Comprising base plate of minimum size 10,000 mm^2, threaded adjustable section including provision for positive location of panels. All pedestals fixed to sub-floor with manufacturer's recommended epoxy adhesive.	To be in accordance with manufacturers stated requirements. Manufacturers lifting device to be used when removing panels.
U1	Unclassified, ie steel and/or aluminium composite adjustable pedestal forming part of system not fully complying or without independent test evidence demonstrating compliance with the PSA MOB Specification.	Unclassified.

For Adjustment factors, Assumptions, Notes, Key failure modes and Key durability issues please see overleaf.

Raised access & platform floors

LOCATIONS - Ground/Intermediate floors

Adjustment Factors

None.

Assumptions

The floor system will be assessed in accordance with the requirements of PSA publication MOB PF2 PS/SPU 1992 (Performance Specification for Platform (Raised Access) Floors) which includes 44 tests to determine the physical design requirements, performance standards and site quality control of the system. Suppliers claiming compliance with this Specification must provide independent test evidence of performance, and state the minimum design life (recommended to be 50 years) for the structure, the panels (recommended to be 25 years) and the maintenance free life of the system, as well as the minimum period of availability of replacement parts. Pedestals form part of a floor system including all necessary accessories, fixings, fastenings and stringers, as well as appropriate panels for the proposed application. Base plate and head to include preformed holes for mechanically fixed applications.

Structural designations within the PSA specification range from light to extra heavy. Typical working areas where the various structural grades may be used are provided as a guide, however, every project should have the appropriate structural load assessed by a competent professional.

- Light – general office accommodation without heavy equipment (permanently marked in green)

- Medium – general office accommodation with some heavy office equipment, data preparation areas, educational accommodation and public areas (permanently marked in yellow)

- Heavy – computer rooms, control rooms (permanently marked in red)

- Extra heavy – computer rooms with heavy equipment and other special applications (permanently marked in blue)

Installation and detailing to comply with both manufacturers guidance and the requirements of PSA Specification. Discrepancies in level of the subfloor must not exceed the standard tolerance allowed by the pedestal support. Where the sub-floor condition, level or strength is unsuitable for adhesive fixing alone of the pedestals (i.e. where there is no test evidence to the PSA site quality control pedestal fixing test), it will be necessary to provide mechanical fixing of pedestals to the floor. The PSA specification requires suppliers to arrange random tests on each project in addition to demonstrate adequacy of fixing without mechanical fixings. Lytag or similar floors will also require mechanical fixing, even if the test is successfully passed, as pedestals are likely to become detached. Heavy items should have spreader plates provided under their feet or supports.

Notes

Copies of the PSA/AFA Performance Specification can be obtained from the Access Flooring Association, who also provide guidance to site installation requirements and use of platform and access flooring systems.

Platform floors are defined as those over a void of 100mm or greater, with a typical maximum of 600mm void. They are suitable for heavy cable applications, including those where the void is used as a plenum for air conditioning. Raised access floors are defined as those over a void less than 100mm, typically the void is 75mm. They are suitable for light cable applications only.

Maintenance (other than cleaning of finishes) should be minimal where the floor panels remain undisturbed. Spreaders etc. to be refitted in their original positions before replacing panels. Pedestals should not be used as "pulleys" for drawing in cables etc. When cleaning surfaces as little water as possible should be used. Islands of panels or pedestals should not be created, particularly when moving furniture etc.

Key failure modes

Instability, unevenness, overloading.

Key durability issues

Condition of sub-floor, loading capacity of system, adequacy of pedestal fixings.

Raised access & _____ platform floors

LOCATIONS - Ground/Intermediate floors

BPG   1 - Flooring Components

Description

General		
	Other pedestals/support systems	**Maintenance**
25	Lightweight concrete block and plastic cap system fully complying with the PSA MOB Specification. Blocks minimum size 10,000 mm² each, and minimum crushing strength 7N/mm 2. Injection moulded ABS caps incorporate locating lugs. Adjustments in floor levels taken up by increments on height of blocks and fine levelling with PVC shims. Blocks fixed to sub-floor with epoxy resin.	To be in accordance with manufacturers stated requirements. Manufacturers lifting device to be used when removing panels.
10	Square pedestals of moisture resistant chipboard, plastic cap and threaded pillar locking screw system fully complying with the PSA MOB Specification. Adjustments in floor levels taken up by packing pieces below cap. Pedestals fixed to sub-floor with epoxy resin, or larger size mechanically fixed.	To be in accordance with manufacturers stated requirements. Manufacturers lifting device to be used when removing panels.
10	PVC mould filled with self-levelling compound screed material. System fully complying with the PSA MOB Specification. Accessibility provided by provision at the time of pouring or by cutting through the screed subsequently.	To be in accordance with manufacturers stated requirements. Manufacturers lifting device to be used when removing panels.
U1	Unclassified, ie pedestal/support forming part of system not fully complying or without independent test evidence demonstrating compliance with the PSA MOB Specification.	Unclassified.

For Adjustment factors, Assumptions, Notes, Key failure modes and Key durability issues please see overleaf.

1.2a

Raised access & platform floors

LOCATIONS - Ground/Intermediate floors

Adjustment Factors

None.

Assumptions

The floor system will be assessed in accordance with the requirements of PSA publication MOB PF2 PS/SPU 1992 (Performance Specification for Platform (Raised Access) Floors) which includes 44 tests to determine the physical design requirements, performance standards and site quality control of the system. Suppliers claiming compliance with this Specification must provide independent test evidence of performance, and state the minimum design life (recommended to be 50 years) for the structure, the panels (recommended to be 25 years) and the maintenance free life of the system, as well as the minimum period of availability of replacement parts. Pedestals form part of a floor system including all necessary accessories, fixings, fastenings and stringers, as well as appropriate panels for the proposed application. Base plate and head to include preformed holes for mechanically fixed applications.

Structural designations within the PSA specification range from light to extra heavy. Typical working areas where the various structural grades may be used are provided as a guide, however, every project should have the appropriate structural load assessed by a competent professional.

- Light   – general office accommodation without heavy equipment (permanently marked in green)

- Medium   – general office accommodation with some heavy office equipment, data preparation areas, educational accommodation and public areas (permanently marked in yellow)

- Heavy   – 7computer rooms, control rooms (permanently marked in red)

- Extra heavy – computer rooms with heavy equipment and other special applications (permanently marked in blue)

Installation and detailing to comply with both manufacturers guidance and the requirements of PSA Specification. Discrepancies in level of the subfloor must not exceed the standard tolerance allowed by the pedestal support. Where the sub-floor condition, level or strength is unsuitable for adhesive fixing alone of the pedestals (i.e. where there is no test evidence to the PSA site quality control pedestal fixing test), it will be necessary to provide mechanical fixing of pedestals to the floor. The PSA specification requires suppliers to arrange random tests on each project in addition to demonstrate adequacy of fixing without mechanical fixings. Lytag or similar floors will also require mechanical fixing, even if the test is successfully passed, as pedestals are likely to become detached. Heavy items should have spreader plates provided under their feet or supports.

Notes

Copies of the PSA/AFA Performance Specification can be obtained from the Access Flooring Association, who also provide guidance to site installation requirements and use of platform and access flooring systems.

Platform floors are defined as those over a void of 100mm or greater, with a typical maximum of 600mm void. They are suitable for heavy cable applications, including those where the void is used as a plenum for air conditioning. Raised access floors are defined as those over a void less than 100mm, typically the void is 75mm. They are suitable for light cable applications only.

Maintenance (other than cleaning of finishes) should be minimal where the floor panels remain undisturbed. Spreaders etc. to be refitted in their original positions before replacing panels. Pedestals should not be used as "pulleys" for drawing in cables etc. When cleaning surfaces as little water as possible should be used. Islands of panels or pedestals should not be created, particularly when moving furniture etc.

Key failure modes

Instability, unevenness, overloading.

Key durability issues

Condition of sub-floor, loading capacity of system, adequacy of pedestal fixings.

Raised access & platform floors

1 - Flooring Components

General	Description	Maintenance
	Floor panels	
35+	Steel tray and top panel, galvanised to BS 1449 Part 1 or BS EN 10142, minimum 275 g/m^2 zinc coating. Central core of mineral fibre / anhydrite / fibre reinforced cement bonded to top panel. Top panel and tray crimped together and composite panel mechanically fixed to pedestals.	To be in accordance with manufacturers stated requirements. Manufacturers lifting device to be used when removing panels.
35+	Aluminium tray and top panel to BS EN 485. Central core of mineral fibre / anhydrite / fibre reinforced cement bonded to top panel. Top panel and tray crimped together and composite panel mechanically fixed to pedestals.	To be in accordance with manufacturers stated requirements. Manufacturers lifting device to be used when removing panels.
30	Steel tray and top panel, galvanised to BS 1449 Part 1 or BS EN 10142, minimum 275 g/m^2 zinc coating. Central core of marine plywood to BS 1088 or M or H grade plywood to BS 6566 Part 8 bonded to top panel. Top panel and tray crimped together and composite panel mechanically fixed to pedestals.	To be in accordance with manufacturers stated requirements. Manufacturers lifting device to be used when removing panels.
25	Steel tray, galvanised to BS 1449 Part 1 or BS EN 10142, minimum 275 g/m^2 zinc coating. Central core of high density particleboard to BS EN 312 grade P5 or P7 bonded to top panel. Top panel and tray crimped together and composite panel mechanically fixed to pedestals.	To be in accordance with manufacturers stated requirements. Manufacturers lifting device to be used when removing panels.
U1	Unclassified, ie composite panel not as described above.	Unclassified.

For Adjustment factors, Assumptions, Notes, Key failure modes and Key durability issues please see overleaf.

Raised access & platform floors

LOCATIONS - Ground/Intermediate floors

Adjustment Factors

Top panel and tray bonded at edge with plastic strip - 5 years

Panel with tray below but no top panel - 5 years

Assumptions

The floor system will be assessed in accordance with the requirements of PSA publication MOB PF2 PS/SPU 1992 (Performance Specification for Platform (Raised Access) Floors) which includes 44 tests to determine the physical design requirements, performance standards and site quality control of the system. Suppliers claiming compliance with this Specification must provide independent test evidence of performance, and state the minimum design life (recommended to be 50 years) for the structure, the panels (recommended to be 25 years) and the maintenance free life of the system, as well as the minimum period of availability of replacement parts. Pedestals form part of a floor system including all necessary accessories, fixings, fastenings and stringers, as well as appropriate panels for the proposed application. Base plate and head to include preformed holes for mechanically fixed applications.

Structural designations within the PSA specification range from light to extra heavy. Typical working areas where the various structural grades may be used are provided as a guide, however, every project should have the appropriate structural load assessed by a competent professional.

- Light   – general office accommodation without heavy equipment (permanently marked in green)

- Medium   – general office accommodation with some heavy office equipment, data preparation areas, educational accommodation and public areas (permanently marked in yellow)

- Heavy   – computer rooms, control rooms (permanently marked in red)

- Extra heavy – computer rooms with heavy equipment and other special applications (permanently marked in blue)

Installation and detailing to comply with both manufacturers guidance and the requirements of PSA Specification. Discrepancies in level of the subfloor must not exceed the standard tolerance allowed by the pedestal support. Where the sub-floor condition, level or strength is unsuitable for adhesive fixing alone of the pedestals (i.e. where there is no test evidence to the PSA site quality control pedestal fixing test), it will be necessary to provide mechanical fixing of pedestals to the floor. The PSA specification requires suppliers to arrange random tests on each project in addition to demonstrate adequacy of fixing without mechanical fixings. Lytag or similar floors will also require mechanical fixing, even if the test is successfully passed, as pedestals are likely to become detached. Heavy items should have spreader plates provided under their feet or supports.

Notes

Copies of the PSA/AFA Performance Specification can be obtained from the Access Flooring Association, who also provide guidance to site installation requirements and use of platform and access flooring systems.

Platform floors are defined as those over a void of 100mm or greater, with a typical maximum of 600mm void. They are suitable for heavy cable applications, including those where the void is used as a plenum for air conditioning. Raised access floors are defined as those over a void less than 100mm, typically the void is 75mm. They are suitable for light cable applications only.

Maintenance (other than cleaning of finishes) should be minimal where the floor panels remain undisturbed. Spreaders etc. to be refitted in their original positions before replacing panels. Pedestals should not be used as "pulleys" for drawing in cables etc. When cleaning surfaces as little water as possible should be used. Islands of panels or pedestals should not be created, particularly when moving furniture etc.

Key failure modes

Instability, unevenness, overloading, deterioration of floor panels (eg corrosion of steel tray, moisture damage to core material).

Key durability issues

Condition of sub-floor, loading capacity of system, adequacy of pedestal fixings.

Entrance matting systems

Scope

This section provides data on entrance matting systems for use in non-domestic building types. It includes matwell frames, plastic and fibrous matting, and a number of different interlocking grid systems comprising metal strips and infill panels, or rubber sections on a wire grid. It should be noted that this section covers only matting materials for use in recessed or surface mounted matwells. Freestanding matting and carpet used as secondary matting, are excluded.

The following component sub-types are included within this section:

	Page
• Matwell frames	1.3
• Interlocking grid/tread systems	1.3
• Interlocking grid systems - replaceable insert strips	1.3a
• Entrance matting	1.3a
• Entrance matting - composite systems	1.3a

The range of different proprietary matting systems currently available to UK specifiers is enormous, and an attempt has been made to categorise the principal types only. Since many systems are designed for specific levels of use or exposure to wetting, it is essential to seek the advices of the manufacturer or supplier prior to making a final selection.

Standards cited

BS 1474: 1987	Specification for wrought aluminium and aluminium alloys for general engineering purposes: bars, extruded round tubes and sections.
BS 5325: 1996	Code of practice for installation of textile floor coverings.
BS 8203: 1996	Code of practice for installation of resilient floor coverings.
BS EN 755: various	Aluminium and aluminium alloys. Extruded rod/bar, tube and profiles.

Other references/information sources

Pye, P. and Harrison, H. 'Floors and flooring - performance, diagnosis, maintenance, repair and the avoidance of defects'. BRE, 1997.

Entrance matting systems

LOCATIONS - Internal, External

1 - Flooring Components

Internal	External	Description	Maintenance
		Matwell frames	
15	15	Extruded aluminium to BS 1474/BS EN 755, epoxy silica coating to exposed treads.	Regular vacuuming and washing down if surface mounted.
15	10	Extruded aluminum to BS 1474/BS EN 755.	Regular vacuuming and washing down if surface mounted.
15	10	Brass.	Regular vacuuming and washing down if surface mounted.
10	10	Co-extruded PVC-u.	Regular vacuuming and washing down if surface mounted.
10	10	Co-extruded acrylic.	Regular vacuuming and washing down if surface mounted.
U1	U1	Unclassified, ie extruded aluminium not to BS 1474/BS EN 755.	Unclassified.
		Interlocking grid/tread systems	
15	15	Extruded aluminium to BS 1474/BS EN 755, epoxy silica coating to exposed treads.	Regular vacuuming and washing down. Replacement of worn/damaged insert strips as necessary.
15	10	Extruded aluminum to BS 1474/BS EN 755.	Regular vacuuming and washing down. Replacement of worn/damaged insert strips as necessary.
15	10	Brass.	Regular vacuuming and washing down. Replacement of worn/damaged insert strips as necessary.
10	10	Co-extruded PVC-u.	Regular vacuuming and washing down. Replacement of worn/damaged insert strips as necessary.
10	10	Co-extruded acrylic.	Regular vacuuming and washing down. Replacement of worn/damaged insert strips as necessary.
U1	U1	Unclassified, ie extruded aluminium not to BS 1474/BS EN 755.	Unclassified.

Entrance matting systems

LOCATIONS - Internal, External

Adjustment factors

Raised/surface mounted matwell frames: -5 years.

Assumptions

There is a vast range of matting systems available, some of which are suitable for use as primary matting (ie heavy duty, internal or external, to remove the bulk of incoming soil),

and some for use as secondary matting only (ie internal use, to remove remaining dirt and moisture). Furthermore, different systems are designed for different levels of trafficking.

The matting system should be selected in accordance with the manufacturer's recommendations, and should be appropriate for the intended location and intensity of use.

Matwell frames may be recessed or surface mounted.

Matwell frame to be securely fixed to base. Matwell base to be level.

Unless otherwise stated by the manufacturer, matting systems for external or excessively wet areas should be of open/drained rather than closed construction.

Installation (including substrate preparation) to be in accordance with manufacturer's instructions, and with BS 5325 (textile floor coverings) or BS 8203 (resilient floor coverings) where relevant.

Excessive use of water during cleaning should be avoided.

Some fading and surface wear may occur within the lives quoted.

Key failure modes

Wear, indentation, scratching, corrosion.

Key durability issues

Location, type/extent of trafficking, selection for intended location/use.

TYPE

Entrance matting systems

LOCATIONS - Internal, External

SUB TYPES
Interlocking grid systems - replaceable insert strips
Entrance matting
Entrance matting - composite systems

BPG | **1 - Flooring Components**

Internal	External	Description	Maintenance
		Interlocking grid systems - replaceable insert strips	
10	10	Solid rubber insert strips.	Regular vacuuming and washing down.
10	5	Fibre reinforced rubber insert strips, ie rubber strips with integral/woven synthetic fibres.	Regular vacuuming and washing down.
10	5	Synthetic fibre brushes (nylon, polyamide) set into synthetic insert strips.	Regular vacuuming, periodic deeper cleaning depending upon intensity of traffic and soiling.
5	3	UV stabilised synthetic fibre (olefin, polypropylene, polyamide, nylon) insert strips.	Regular vacuuming, periodic deeper cleaning depending upon intensity of traffic and soiling.
5	U1	Non-UV stabilised synthetic fibre (olefin, polypropylene, polyamide, nylon) insert strips.	Regular vacuuming, periodic deeper cleaning depending upon intensity of traffic and soiling.
3	U2	Natural fibre (coir) inset strips.	Regular vacuuming, periodic deeper cleaning depending upon intensity of traffic and soiling.
		Entrance matting	
10	10	Rubber (solid, grid).	Regular vacuuming and washing down.
5	5	UV resistant PVC grid.	Regular vacuuming and washing down.
5	U1	Natural fibre (coir).	Regular vacuuming, periodic deeper cleaning depending upon intensity of traffic and soiling.
3	U2	PVC backed synthetic fibre (polyamide, polypropylene).	Regular vacuuming, periodic deeper cleaning depending upon intensity of traffic and soiling.
		Entrance matting - composite systems	
10	10	Rubber/neoprene rubber/EPDM sections/strips, with/without aluminium/brass/PVC-u spacers, jointed together with high tensile austenitic stainless steel or aluminium wire.	Regular vacuuming, periodic deeper cleaning depending upon intensity of traffic and soiling.
10	5	Fibre reinforced rubber sections/strips, with/without aluminium/brass/PVC-u spacers, jointed together with high tensile austenitic stainless steel or aluminum wire.	Regular vacuuming, periodic deeper cleaning depending upon intensity of traffic and soiling.
10	5	Synthetic fibre brushes (nylon, polyamide) set into synthetic insert strips, jointed together with high tensile austenitic stainless steel or aluminium wire.	Regular vacuuming and washing down.
5	5	Rubber/neoprene rubber/EPDM/fibre reinforced rubber sections/strips, with/without aluminium/brass/PVC-u spacers, jointed together with high tensile galvanised steel wire.	Regular vacuuming, periodic deeper cleaning depending upon intensity of traffic and soiling.
5	5	Synthetic fibre brushes (nylon, polyamide) set into synthetic insert strips, jointed together with high tensile galvanised steel wire.	Regular vacuuming and washing down.
5	U1	Rubberised polyester strips, set between aluminum or PVC-u sections, jointed together with high tensile galvanised steel wire.	Regular vacuuming and washing down.

1.3a | For **Adjustment factors, Assumptions, Key failure modes and Key durability issues please see overleaf.**

Entrance matting systems

LOCATIONS - Internal, External

Adjustment factors

Matting/insert strips subject to severe abrasive conditions, eg traffic by loaded vehicles with steel or rigid plastic wheels: -3 years.

Use in lightly trafficked areas, eg secondary entrances, single family dwellings: +2 years.

Assumptions

Matwell frames may be recessed or surface mounted.

There is a vast range of matting systems available, some of which are suitable for use as primary matting (ie heavy duty, internal or external, to remove the bulk of incoming soil),

and some for use as secondary matting only (ie internal use, to remove remaining dirt and moisture). Furthermore, different systems are designed for different levels of trafficking.

The matting system should be selected in accordance with the manufacturer's recommendations, and should be appropriate for the intended location and intensity of use.

Unless otherwise stated by the manufacturer, matting systems for external of excessively wet areas should be of open/drained rather than closed construction.

Installation (including substrate preparation) to be in accordance with manufacturer's instructions, and with BS 5325 (textile floor coverings) or BS 8203 (resilient floor coverings) where relevant.

Some fading and surface wear may occur within the lives quoted.

Key failure modes

Corrosion of jointing wires, rotting/decay of fibres, UV degradation of synthetic fibres, wear, loss of fibres, surface scratching.

Key durability issues

Type/intensity of use, pile thickness, specification of frame/scraper bars, jointing wires, selection for intended use.

1.3a

LOCATIONS - General

1 - Flooring Components

General	Description	Maintenance
	Flexible PVC sheet	
15	Flexible PVC sheet flooring to BS EN 649, minimum use class 33 to BS EN 685 (heavy commercial). Minimum 4mm overall thickness, with glass-fibre reinforcement, aluminium oxide/quartz/silicon carbide granules throughout the thickness of the material, and polyester/cellulose backing. Wear surface dressed with silicon carbide/quartz granules or grit. BBA or other 3rd party assurance for the proposed use.	Regular sweeping/vacuuming, wet mopping or periodic machine scrubbing with mild, neutral detergent, in accordance with manufacturer's instructions and/or BS 6263:Part 2. Polish only if in accordance with manufacturer's directions.
10	Flexible PVC sheet flooring to BS EN 649, minimum use class 33 to BS EN 685 (heavy commercial). Minimum 2mm overall thickness, with glass-fibre reinforcement, aluminium oxide/quartz/silicon carbide granules throughout the thickness of the material, and polyester/cellulose backing. Wear surface dressed with silicon carbide/quartz granules or grit. BBA or other 3rd party assurance for the proposed use.	Regular sweeping/vacuuming, wet mopping or periodic machine scrubbing with mild, neutral detergent, in accordance with manufacturer's instructions and/or BS 6263:Part 2. Polish only if in accordance with manufacturer's directions.
10	Flexible PVC sheet flooring to BS EN 649, minimum use class 32 to BS EN 685 (general commercial). Minimum 2mm overall thickness, with aluminium oxide/quartz/silicon carbide granules throughout the thickness of the material. BBA or other 3rd party assurance for the proposed use.	Regular sweeping/vacuuming, wet mopping or periodic machine scrubbing with mild, neutral detergent, in accordance with manufacturer's instructions and/or BS 6263:Part 2. Polish only if in accordance with manufacturer's directions.
10	Flexible PVC sheet flooring to BS EN 649, minimum use class 32 to BS EN 695 (general commercial). Minimum 2mm overall thickness, with wear surfaces dressed with silicon carbide/quartz granules or grit. BBA or other 3rd party assurance for the proposed use.	Regular sweeping/vacuuming, wet mopping or periodic machine scrubbing with mild, neutral detergent, in accordance with manufacturer's instructions and/or BS 6263:Part 2. Polish only if in accordance with manufacturer's directions.
10	Flexible PVC sheet flooring to BS EN 649, minimum 2mm overall thickness. Wear surfaces dressed with PVC and cork chips. Use restricted to BS EN 685 class 31 (light commercial). BBA or other 3rd party assurance for the proposed use.	Regular sweeping/vacuuming, wet mopping or periodic machine scrubbing with mild, neutral detergent, in accordance with manufacturer's instructions and/or BS 6263:Part 2. Polish only if in accordance with manufacturer's directions.
U1	Unclassified, ie PVC sheet flooring not to BS EN 649 and/or specification/use class not specified.	Unclassified.

For Adjustment factors, Assumptions, Key failure modes and Key durability issues please see overleaf.

Non-slip floor coverings

LOCATIONS - General

Key failure modes

Surface wear/deterioration, cracking, bubbles/ridges/ rippling, indentation, chemical attack, UV degradation, shrinkage.

Key durability issues

Preparation/condition of base (especially flatness, effectiveness of DPM), adequate/correct maintenance, suitability for use/location, thickness of sheeting.

Assumptions

The appropriate grade of flooring must be selected for the intended location/intensity of use, in accordance with the classification system in BS EN 685, which defines ten different

levels of use covering domestic, commercial and light industrial applications. An alternative classification system, used in Agreement Certificates, is provided in UEAtc M.O.A.T. No. 36.

Installation (including preparation/condition of base) to be in accordance with BS 8203 and manufacturer's instructions. Flatness of the base is critical. Floor covering fully bonded to sub-floor using an adhesive approved by manufacturer. All joints to be hot welded.

Ground bearing solid slabs to have adequate damp proof membrane in accordance with CP 102/BS 8102. Any expansion joints in the base should be carried through the flooring by way of a proprietary joint. Excessive use of water during cleaning should be avoided.

Protective nosings to be fitted if flooring used on staircases/landings.

Some flooring materials may be permanently marked by concentrated acids, organic solvents and certain dyes.

Adjustment factors

Flooring thickness of 3.5mm or greater: +2 years.

Light industrial use: life limited to 10 years.

Flooring without BBA or other 3rd party assurance for proposed use: -2 years.

1.4

Walling and Cladding Components

BPG

Walling and Cladding Components

Structural frames

Scope

This section provides data on structural framing materials for use in the walls, floors and roofs of non-domestic building types. It includes hot rolled and cold formed steel, precast and insitu concrete, and glued-laminated timber members. Other timber members are excluded from this section, but are included in Sections 1-3 of the HAPM Component Life Manual. Proprietary structural framing systems are also excluded.

The following component sub-types are included within this section:

	Page
• Hot rolled steel	2.1
• Cold formed steel	2.1a
• Concrete encased steel	2.1b
• Reinforced dense concrete	2.1c
• Pre-stressed precast concrete	2.1c
• Glued laminated timber	2.1d

It should be noted that this section is structured according to the principal materials and, where relevant, types of surface protection available in the UK. Specific consideration of proprietary structural framing systems is outside the scope of this section.

Standards cited

BS 729: 1971(1994)	Specification for hot dip galvanized coatings on iron and steel articles.
BS 970:	Specification for wrought steels for mechanical and allied engineering purposes.
Part 1: 1996	General inspection and testing procedures and specific requirements for carbon, carbon manganese, alloy and stainless steels.
BS 1204:1993	Specification for type MR phenolic and aminoplastic synthetic resin adhesives for wood.
BS1449:	Steel plate, sheet and strip.
Part 1:1991	Carbon and carbon manganese plate, sheet and strip.
BS 2994:1976(1987)	Specification for cold rolled steel sections.
BS 4169:1988	Specification for manufacture of glued-laminated timber structural members.
BS 4232:1967(withdrawn)	Specification for surface finish of blast-cleaned steel for painting.
BS 4449:1997	Specification for carbon steel bars for the reinforcement of concrete.
BS 4848:various	Hot rolled structural steel sections.
BS 4978:1996	Specification for visual strength grading of softwood.
BS 5268:	Structural use of timber.
Part 5:1989 (1997)	Code of practice for the preservative treatment of structural timber.
BS 5328:	Concrete.
Part 1: 1997	Guide to specifying concrete.
Part 2: 1997	Methods for specifying concrete mixers.
Part 3: 1990	Specification for the procedures to be used in producing and transporting concrete.
Part 4: 1990	Specification for the procedures to be used in sampling, testing and assessing compliance of concrete.
BS 5493: 1977	Code of practice for protective coating of iron and steel structures against erosion.
BS 8110:	Structural use of concrete.
Part 1: 1997	Code of practice for design and construction.
Part 2: 1985	Code of practice for special circumstances.
BS PD 6484:1979(1990)	Commentary on corrosion at bimetallic contacts and its alleviation.
BS EN 204:1991	Classification of non-structural adhesives for joining of wood and derived timber products.

Continued overleaf

Structural frames

Standards cited *continued*

BS EN 301:1992	Adhesives, phenolic and aminoplastic, for load-bearing timber structures: classification and performance requirements.
BS EN 350:	Durability of wood and wood-based products. Natural durability of solid wood.
Part 2:1994	Guide to natural durability and treatability of natural wood species of importance in Europe.
BS EN 386:1995	Glued laminated timber. Performance requirements and minimum production requirements.
BS EN 519:1995	Structural timber. Grading. Requirements for machine strength graded timber and grading machines.
BS EN 10142:1991	Specification for continuously hot-dip zinc coated low carbon steel sheet and strip for cold forming: technical delivery conditions.
BS EN 10147:1992	Specification for continuously hot-dip zinc coated structural steel sheet and strip. Technical delivery conditions.

Other references/information sources

Galvanizers' Association: Engineers and Architects' guide to hot dip galvanizing.

British Steel: 'The prevention of corrosion on Structural Steelwork'.

TRADA Wood Information Sheets.

Glued Laminated Timber Association specifiers' guide.

Structural frames

LOCATIONS - External, Internal

BPG   2 - Walling & Cladding Components

External	Internal	Description	Maintenance
		Hot rolled steel	
30	35+	Mild steel, hot dipped galvanized after any cutting, welding or drilling to BS 5493 specification SB3 (equivalent in either case to a zinc coating weight of 1500g/m^2 or 210 micron thickness).	None.
25	35+	Mild steel, hot dipped galvanized after any cutting, welding or drilling to BS 5493 specification SB2 (equivalent in either case to a zinc coating weight of 1000g/m^2 or 140 micron thickness).	None.
25	35+	Mild steel, molten zinc sprayed after any cutting, welding or drilling to BS 5493 specification SC2Z (equivalent to a zinc coating weight of about 1050g/m^2 or 150 micron thickness).	None.
20	35+	Mild steel, molten zinc sprayed after any cutting, welding or drilling to BS 5493 specification SC1Z (equivalent to a zinc coating weight of about 700g/m^2 or 100 micron thickness).	None.
20	35	Mild steel, hot dipped galvanized after any cutting, welding or drilling to BS 729 (to steel at least 5mm thick) or to BS 5493 specification SB1 (any thickness steel) equivalent in either case to a zinc coating weight of 610g/m^2 or 85 micron thickness.	None.
15	35	Mild steel, hot dipped galvanized after any cutting, welding or drilling to BS 729 to steel 2mm - 4.9mm thick (equivalent to a zinc coating weight of 460g/m^2 or 64 micron thickness).	None.
10	30	Mild steel, hot dipped galvanized after any cutting, welding or drilling to BS 729 to steel 1 - 1.9mm thick (equivalent to a zinc coating weight of 335g/m^2 or 43 micron thickness).	None.
10	25	Mild steel, blast cleaned after any cutting, welding or drilling to BS 4232 'second quality' (equivalent to SA 2.5) and protected with an appropriate Micaceous Iron Oxide, Chlorinated Rubber or similar high performance finish to give a minimum dry film thickness of 250 microns.	None.
U1	20	Mild steel, protected with two generous coats of Bitumen Solution Paint (ie solvent based bitumen, NOT the more widely available Bitumen Emulsions).	None.
U2	U1	Unclassified, ie mild steel 'factory primed'.	Unclassified.

For Adjustment factors, Assumptions, Notes, Key failure modes and Key durability issues please see overleaf.

2.1

Structural frames

LOCATIONS - External, Internal

Adjustment factors

Industrial/polluted/marine environments: -5 years.

Assumptions

These life assessments represent the life to first maintenance. Whether maintenance is possible or even probable will depend upon the feasibility of access to ALL surfaces. In most cases such access will not be possible.

Steel to BS 1449:Part 1 or BS 970:Part 1 and BS 4848 and BS 2994 sections.

Surface preparation and coating application in accordance with BS 5493 and/or BS 729.

Steel frame is assumed to be subject to the following environments:

– External: an environment that would be classified as 'exterior exposed non-polluted inland' using Table 3 Part 1 to BS 5493.

– Internal: an environment that would be classed as 'interior, normally dry' using Table 3 Part 6 of BS 5493, and which will not be exposed to humidity in excess of 70% for significant periods.

For more severe environments, reference should be made to the relevant tables in BS 5493.

Structural members concealed within the external envelope to be in well-ventilated cavities with an effective vapour control layer on the indoor side and suitable detailing to avoid wetting by rainwater or condensate.

Frames to be connected to an earth point common to the electrical services.

Notes

Weights of zinc coatings are for any one face only.

The range of options for protecting steel is huge, therefore the classes listed are indicative only.

The lives and classes included are intended for solid sections only; the complications of protecting hollow sections (to avoid internal corrosion) have been ignored for simplicity.

Key failure modes

Corrosion, chalking/peeling of surface finish, impact.

Key durability issues

Type/thickness of surface protection, exposure conditions.

Structural frames

BPG | 2 - Walling & Cladding Components

LOCATIONS - Internal

Internal	Description	Maintenance
	Cold formed steel	
35+	Austenitic stainless steel to BS 970:Part 1, grade 316 or 304.	None.
35+	Post-galvanized mild steel, minimum 600g/m² zinc coating weight.	If painted, redecorate every 5 years.
35+	Pre-galvanized mild steel, minimum 600g/m² zinc coating weight. Factory applied organic coating, 25-50 microns nominal thickness.	Coatings may require redecoration after 10-15 years.
35+	Mild steel, molten zinc sprayed after fabrication, minimum 700g/m² zinc coating weight.	If painted, redecorate every 5 years.
35	Pre-galvanized or post-galvanized mild steel, minimum 450g/m² zinc coating weight. Factory applied organic coating, 25-50 microns nominal thickness.	Coatings may require redecoration after 10-15 years.
30	Pre-galvanized or post-galvanized mild steel, minimum 275g/m² zinc coating weight. Factory applied organic coating, 25-50 microns nominal thickness.	Coatings may require redecoration after 10-15 years.
25	Pre-galvanized or post-galvanized mild steel, minimum 275g/m² zinc coating weight.	If painted, redecorate every 5 years.
25	Mild steel protected with an appropriate Micaceous Iron Oxide, Chlorinated Rubber or similar high performance finish to give a minimum dry film thickness of 250 microns.	Redecorate every 5 years.
U1	Unclassified, ie mild steel 'factory primed' or less than above specifications.	Unclassified.

For Adjustment factors, Assumptions, Notes, Key failure modes and Key durability issues please see overleaf.

2.1a

Structural frames

LOCATIONS - Internal

Notes

The range of options for protecting steel is huge, therefore the classes listed are indicative only.

Key failure modes

Corrosion, chalking/peeling of surface finish, impact/indentation/distortion.

Key durability issues

Type/thickness of surface protection, exposure conditions, maintenance frequency.

Structural members concealed within the external envelope to be in well-ventilated cavities with an effective vapour control layer on the indoor side and suitable detailing to avoid wetting by rainwater or condensate.

Any welded connections to be subsequently cleaned and painted with zinc rich paint to provide a minimum film thickness of 50 microns.

Frames to be connected to an earth point common to the electrical services.

With galvanized hollow sections it is essential to provide holes for venting and drainage, and to ensure that the internal surfaces are fully coated.

Where two or more metals are used in the frame construction, they should be compatible, ie to prevent galvanic corrosion. For further detailed guidance, see BS PD 6484.

Adjustment factors

Through dip in bitumen: +5 years.

Assumptions

These life assessments represent the life to first maintenance. Whether maintenance is possible or even probable will depend upon the feasibility of access to ALL surfaces. In most cases such access will not be possible.

Mild steel to BS EN 10147 (post-galvanized sheet), BS EN 10142 (pre-galvanized sheet), BS 1449:Part 1.

Surface preparation and coating application in accordance with BS 5493 and/or BS 729.

Steel frame is assumed to be subject to an environment that would be classed as 'interior, normally dry' using Table 3 Part 6 of BS 5493, and which will not be exposed to humidity in excess of 70% for significant periods. For more severe environments, reference should be made to the relevant table in BS 5493.

2.1a

Structural frames _____

BPG 2 - Walling & Cladding Components

Internal	Description	Maintenance
	Concrete encased steel	
35+	Hot rolled steel encased in insitu concrete to BS 8110:Parts 1and 2, with cover to steel complying with Table 3.3 and specified to match the appropriate exposure category selected from Table 3.2. Minimum nominal cover to all reinforcement of 20mm. Minimum cement content 275kg/m³.	None.
U1	Unclassified, ie concrete encased steel beams, concrete encasement not complying with BS 8110:Parts 1 and 2.	Unclassified.

Adjustment factors

Structural members subjected to 'very severe' exposure conditions to Table 3.2 of BS 8110:Part 1, without additional surface protection: -10 years.

Assumptions

Concrete specified in accordance with BS 5328:Parts 1-4.

Steel to BS 1449:Part 1 or BS 970:Part 1 and BS 4848 and BS 2994 sections.

Steel protected with factory applied primer coat.

Structural members will not be subjected to the exposure class designated 'most severe' in Table 3.2 of BS 8110:Part 1 (frequent exposure to sea water spray or de-icing salts, immersion in sea water).

Concrete structural members in ground contact to be completely protected from contact with sulphate salts in soils classified as Class 5 in BS 5328:Part 1. For other soil sulphate classes, minimum cement content to comply with Table 7 of BS 5328:Part 1.

Surface protection (eg paint films, membranes) should be applied to structural members subjected to 'very severe' exposure conditions to Table 3.2 of BS 8110:Part 1.

Detailing to ensure free drainage from exposed horizontal surfaces which are not otherwise protected from the weather.

Notes

None.

Key failure modes

Corrosion of steelwork (eg due to carbonation, sulphate/chloride attack), spalling, cracking/crazing.

Key durability issues

Thickness of concrete encasement, curing/compaction of cement mix, cement content suitable for exposure, protection against sulphates, chlorides, acids (eg fumes, chemical spills), detailing.

2.1b

LOCATIONS - General

BPG | **2 - Walling & Cladding Components**

General	Description	Maintenance
	Reinforced dense concrete	
35+	Insitu columns/beams complying with BS 8110:Parts 1and 2, with cover to steel complying with Table 3.3 and specified to match the appropriate exposure category selected from Table 3.2. Minimum nominal cover to all reinforcement of 20mm. Minimum cement content 275kg/m³.	None.
35+	Precast columns/beams complying with BS 8110:Parts 1and 2, with cover to steel complying with Table 3.3 and specified to match the appropriate exposure category selected from Table 3.2. Minimum nominal cover to all reinforcement of 20mm. Minimum cement content 275kg/m³.	None.
U1	Unclassified, ie beams/columns not complying with BS 8110:Parts 1 and 2.	Unclassified.
	Pre-stressed precast concrete	
35+	Columns/beams complying with BS 8110:Parts 1and 2, with cover to steel complying with Table 4.8 and specified to match the appropriate exposure category selected from Table 3.2. Minimum nominal cover to all reinforcement of 20mm. Minimum cement content 300kg/m³.	None.
U1	Unclassified, ie beams/columns not complying with BS 8110:Parts 1 and 2.	Unclassified.

For Adjustment factors, Assumptions, Notes, Key failure modes and Key durability issues please see overleaf.

2.1c

Structural frames

Adjustment factors

Structural members subjected to 'very severe' exposure conditions to Table 3.2 of BS 8110:Part 1, without additional surface protection: -10 years.

Assumptions

Concrete specified in accordance with BS 5328:Parts 1-4.

Mild steel reinforcement to BS 4449 (reinforced dense concrete only).

Design of structural connections to precast members to be in accordance with BS 8110:Parts 1 and 2. Connections should be designed to maintain the level of protection against weather, fire and corrosion required for the remainder of the structure.

Structural members will not be subjected to the exposure class designated 'most severe' in Table 3.2 of BS 8110:Part 1 (frequent exposure to sea water spray or de-icing salts, immersion in sea water).

Concrete structural members in ground contact to be completely protected from contact with sulphate salts in soils classified as Class 5 in BS 5328:Part 1. For other soil sulphate classes, minimum cement content to comply with Table 7 of BS 5328:Part 1.

Surface protection (eg paint films, membranes) should be applied to structural members subjected to 'very severe' exposure conditions to Table 3.2 of BS 8110:Part 1.

Detailing to ensure free drainage from exposed horizontal surfaces which are not otherwise protected from the weather.

Precast members with mechanical damage or cracking must not be installed. Precast beams must have adequate bearing, in accordance with manufacturers' directions. Beams should be evenly packed under with mortar at their bearings. The ends of beams supported on the inner leaf of cavity walls must not encroach into the cavity, as this risks both bridging of the cavity and repeated wetting of the beam.

Notes

Overloading of beams must be avoided. The most common causes of overloading are point-loading or poor handling during storage and installation, or due to changes in specification (eg from a floating deck to a screed, or from lightweight partitions to masonry partitions). Such changes must be advised to the beam manufacturer or the structural calculations rechecked for adequacy.

For pre-stressed wide planks, see HAPM Component Life Manual p.1.3.

Key failure modes

Reinforcement corrosion (eg due to carbonation, sulphate/ chloride attack), spalling, cracking/crazing.

Key durability issues

Thickness of reinforcement cover, curing/compaction of cement mix, cement content suitable for exposure, protection against sulphates, chlorides, acids (eg fumes, chemical spills), detailing.

Structural frames

BPG 2 - Walling & Cladding Components

External	Internal	Description	Maintenance
		Glued laminated timber	
35	35+	Glued laminated members manufactured to BS EN 386 from timber of Class 1 ('very durable') to BS EN 350:Part 2, using Type 1 adhesive to BS EN 301.	Redecorate: stain every 3 years or paint/varnish every 5 years (external), stain every 5 years, paint/varnish every 10 years (internal).
30	35+	Glued laminated members manufactured to BS EN 386 from timber of Class 2 ('durable') to BS EN 350:Part 2, using Type 1 adhesive to BS EN 301.	Redecorate: stain every 3 years or paint/varnish every 5 years (external), stain every 5 years, paint/varnish every 10 years (internal).
30	35+	Glued laminated members manufactured to BS EN 386 from European Redwood using Type 1 adhesive to BS EN 301. Double vacuum impregnated after machining with organic solvent preservative to the appropriate treatment schedule specified in BS 5268:Part 5.	Redecorate: stain every 3 years or paint/varnish every 5 years (external), stain every 5 years, paint/varnish every 10 years (internal).
25	35+	Glued laminated members manufactured to BS EN 386 from European Whitewood using Type 1 adhesive to BS EN 301. Double vacuum impregnated after machining with organic solvent preservative to the appropriate treatment schedule specified in BS 5268:Part 5.	Redecorate: stain every 3 years or paint/varnish every 5 years (external), stain every 5 years, paint/varnish every 10 years (internal).
10	35+	Glued laminated members manufactured to BS EN 386 from European Redwood or whitewood using Type 1 adhesive to BS EN 301. Immersion treated after machining with organic solvent preservative.	Redecorate: stain every 3 years or paint/varnish every 5 years (external), stain every 5 years, paint/varnish every 10 years (internal).
U1	35	Glued laminated members manufactured to BS EN 386 using Type II adhesive to BS EN 301.	Redecorate: stain every 3 years or paint/varnish every 5 years (external), stain every 5 years, paint/varnish every 10 years (internal).
U2	35	'Standard Grade' glued laminated members manufactured to BS 4169 using 'WBP' adhesive to BS 1204:Part 1 ('D4' adhesive to BS EN 204). Available from manufacturers as stock items.	Redecorate: stain every 3 years or paint/varnish every 5 years (external), stain every 5 years, paint/varnish every 10 years (internal).
U3	U1	Unclassified, ie glued laminated members not to BS EN 386 or BS 4169.	Unclassified.

For Adjustment factors, Assumptions, Notes, Key failure modes and Key durability issues please see overleaf.

2.1d

Structural frames

LOCATIONS - Internal, External

Adjustment factors

Internal use in service conditions in which the moisture content of the member is expected to exceed 20% for more than three weeks per year (eg unventilated or poorly ventilated voids, use within swimming pools, laundries etc): -10 years.

Assumptions

BS EN 301 Type 1 adhesive to be used where members exposed to the weather or subject to prolonged high temperatures (NB: in some roofs, temperatures can exceed 50 deg C during summer).

Where the moisture content of the member in service is likely to exceed 20%, either the timber should be naturally resistant to fungal decay, or preservative treatment will be required.

Internal timbers may require protection against insect attack.

Organic solvent preservatives to be to BWPA Table 4 Type F/N (ie Fungicidal and Insecticidal).

All timber included in roofs of buildings constructed in the areas listed in the Approved Document to support Regulation 7 of the Building Regulations (ie areas of Surrey, Berkshire and Hampshire which are liable to attack by the House Longhorn Beetle - Hylotrupes Bajulus) must be adequately treated to resist insect attack.

Timber preservatives or other surface treatments (eg flame retardants) must be compatible with adhesive.

Timber laminates to be strength graded to BS 4978, BS EN 519 (or other acceptable grading standard).

Fixings or connectors to be plated or non-ferrous.

Direct contact with the ground to be avoided.

Notes

BS 4169 has been partly superceded by BS EN 386. However, 'stock items' to BS 4978 are still obtainable.

Glued laminated members produced to BS EN 386 should be marked with the manufacturer's identity, strength class, adhesive type, date of production, third party quality control certificate number and the standard number. Similar requirements apply for BS 4169.

Key failure modes

Fungal/insect attack, cracks/splits, delamination.

Key durability issues

Timber species and permeability, preservative treatment, surface protection, maintenance frequency, adhesive type, exposure conditions.

2.1d

Profiled metal claddings

Scope

This section provides data on profiled metal cladding systems for use in the walls and roofs of non-domestic building types. It includes stainless steel and mild steel, coated and uncoated aluminium, and a range of common surface protection systems.

The following component sub-types are included within this section:

	Page
• Stainless steel	2.2
• Mild steel	2.2a
• Aluminium	2.2b

Metal panels for installation between metal framing members are included in Section 2.6 of this manual. Thermal insulation materials are included in Sections 2 and 3 of the HAPM Component Life Manual. Sealants and gaskets are included in Section 4, and plastic and fibre cement profiled claddings in Section 2, of the HAPM Manual.

Standards cited

BS 970:

Part 1:1996 — Specification for wrought steels for mechanical and allied engineering purposes.

General inspection and testing procedures and specific requirements for carbon, carbon manganese, alloy and stainless steels.

BS 1449:

Part 2:1983 — Steel plate, sheet and strip.

Specification for stainless and heat resisting steel plate, sheet and strip.

BS 1474:1987 — Specification for wrought aluminium and aluminium alloys for general engineering purposes: bars, extruded round tubes and sections.

BS 1494:

Part 1:1964 — Specification for fixing accessories for building purposes.

Fixings for sheet, roof and wall coverings.

BS 5427:

Part 1:1996 — Code of practice for the use of profiled sheet for roof and wall cladding on buildings.

Design.

BS 6582:1985 — Specification for continuously hot-dip lead alloy (terne) coated cold reduced carbon steel flat rolled products.

BS 8200:1985 — Code of practice for design of non-loadbearing external vertical enclosures of buildings.

BS PD 6484:1979(1990) — Commentary on corrosion at bimetallic contacts and its alleviation.

BS EN 485: various — Aluminium and aluminium alloys. Sheet, strip and plate.

BS EN 10088:

Part 2:1995 — Stainless steels.

Technical delivery conditions for sheet/plate and strip for general purposes.

BS EN 10142:1991 — Specification for continuously hot-dip zinc coated low carbon steel sheet and strip for cold forming: technical delivery conditions.

BS EN 10147:1992 — Specification for continuously hot-dip zinc coated structural steel sheet and strip. Technical delivery conditions.

BS EN 10215:1995 — Continuously hot-dip zinc-aluminium (AZ) coated steel strip and sheet. Technical delivery conditions.

Other references/information sources

Metal Cladding & Roofing Manufacturers' Association Design Guides.

National Federation of Roofing Contractors: 'Profiled sheet metal roofing and cladding - a guide to good practice'.

Ryan, P. et al. 'Durability of Cladding - a state of the art report'. Thomas Telford, 1994.

Oliver, M. et al. 'Coated metal roofing and cladding'. BBA, 1997.

Profiled metal claddings

LOCATIONS - Walls, Roofs

2 - Walling & Cladding Components

General	Description	Maintenance
	Stainless steel	
35+	Austenitic stainless steel to BS 1449:Part 2 or BS 970:Part 1, grade 316, or BS EN 10088-2, grade 1.4401. Terne coated to BS 6582.	Annual inspection & maintenance in accordance with BS 5427:Part 1, Table 9. Renewal of side/end lap seals as necessary.
35+	Austenitic stainless steel to BS 1449:Part 2 or BS 970:Part 1, grade 316, or BS EN 10088-2, grade 1.4401.	Annual inspection & maintenance in accordance with BS 5427:Part 1, Table 9. Regular cleaning with non-alkaline detergent (polluted/marine areas every 3 months, other areas every 6 months). Renewal of side/end lap seals as necessary.
30	Austenitic stainless steel to BS 1449:Part 2 or BS 970:Part 1, grade 304, or BS EN 10088-2, grade 1.4301. Terne coated to BS 6582. Not suitable for use in exposed marine or aggressive industrial atmospheres.	Annual inspection & maintenance in accordance with BS 5427:Part 1, Table 9. Renewal of side/end lap seals as necessary.
30	Austenitic stainless steel to BS 1449:Part 2 or BS 970:Part 1, grade 304, or BS EN 10088-2, grade 1.4301. Factory applied organic coating, 25-50 microns nominal thickness. Not suitable for use in exposed marine or aggressive industrial atmospheres.	Annual inspection & maintenance in accordance with BS 5427:Part 1, Table 9. Regular cleaning with non-alkaline detergent (polluted/marine areas every 3 months, other areas every 6 months). Renewal of side/end lap seals as necessary.
25	Austenitic stainless steel to BS 1449:Part 2 or BS 970:Part 1, grade 304, or BS EN 10088-2, grade 1.4301. Not suitable for use in exposed marine or aggressive industrial atmospheres.	Annual inspection & maintenance in accordance with BS 5427:Part 1, Table 9. Regular cleaning with non-alkaline detergent (polluted/marine areas every 3 months, other areas every 6 months). Renewal of side/end lap seals as necessary.
20	Ferritic stainless steel to BS 1449:Part 2 or BS 970:Part 1, grade 430 or 434, or BS EN 10088-2, grade 1.4000 or 1.4113. Factory applied organic coating, 25-50 microns nominal thickness. For use with manufacturer's approval in mild atmospheres only.	Annual inspection & maintenance in accordance with BS 5427:Part 1, Table 9. Regular cleaning with non-alkaline detergent (polluted/marine areas every 3 months, other areas every 6 months). Renewal of side/end lap seals as necessary.
U1	Unclassified, ie stainless steel not to BS 1449:Part 2, BS 970:Part 1 or BS EN 10088-2, or less than above specifications.	Unclassified.

For Adjustment factors, Assumptions, Notes, Key failure modes and Key durability issues please see overleaf.

Profiled metal claddings

LOCATIONS - Walls, Roofs

Adjustment factors

Use in industrial/polluted/marine environment: -5 years (except 316 grade austenitic stainless steel).

Galvanized/plated steel fixings: life limited to 25 years.

Assumptions

Surface coatings may deteriorate during the lives quoted. Recoating is feasible but is not normally practical and is rarely carried out.

Profiled metal cladding may be single skin, double skin with profiled metal liner, insulation and outer skin assembled on site, or factory made composite panels with a rigid foam core.

Design and installation in accordance with BS 5427: Part 1, BS 8200 and in strict accordance with manufacturer's instructions.

Adequate provision for movement to be made in the system when installed.

Fixings to be plated carbon steel or stainless steel to BS 1494:Part 1, or proprietary fixings in accordance with manufacturer's guidance, and to be compatible with other metals used.

Type and spacing of fixing in strict accordance with manufacturer's directions. Heads of plated steel fixings to be protected, eg by push-on caps or integral plastic heads.

Spacing of supports and spacer bars in accordance with manufacturer's load-span tables. Spacer bars, channels, supporting purlins etc to be minimum $275g/m^2$ galvanized steel or no less durable than the cladding.

Provision of vapour control layer, breather membrane and ventilation to built up systems in accordance with the guidance in BS 5427:Part 1.

Sealants to side and end laps (where required by BS 5427:Part 1) to be non-setting gap filling, or pre-formed tape, and compatible with the metal and its coating, and UV resistant if exposed to sunlight.

Provision of movement joints in accordance with BS 5427:Part 1, Table 5.

Coating of reverse side of outer sheet and of liner panels to be appropriate for the environment. Liner panels to be minimum $275g/m^2$ galvanized mild steel, with minimum 20 micron organic coating.

Where two or more metals are used in the cladding construction, they should be compatible or isolated from each other, ie to prevent galvanic corrosion. For further detailed guidance, see BS PD 6484.

Where the roof is designed for regular access, special walkways should be provided to prevent damage to the cladding.

Notes

The life of metal coatings is influenced by the colour of the coating, the orientation of the building, and by the environmental conditions. Detailed guidance on lives and on any guarantee or warranty schemes is available from manufacturers. Proprietary products are available for the re-coating of factory applied coatings.

Corrosion of unprotected cut edges is a common problem which can occur within a relatively short space of time. This process can be delayed by the application of additional edge protection at the time of installation (eg latex system or other products recommended by sheet supplier).

Terne coatings are applied for decorative purposes only.

Some fading and chalking of coatings may occur within the life quoted.

Further guidance on the design and installation of profiled metal cladding is available from the Metal Cladding and Roofing Manufacturers' Association and the National Federation of Roofing Contractors.

Key failure modes

Corrosion (particularly at cut edges, fixings and changes in profile), scratching/chalking/blistering/peeling of surface coating, impact/indentation/distortion, elongation of fixing holes, pull-out of fixings, debonding of skin and core (composite panels).

Key durability issues

Base metal, type/thickness of surface protection, maintenance frequency, exposure conditions, colour of coating.

Profiled metal claddings

LOCATIONS - Walls, Roofs

BPG 2 - Walling & Cladding Components

General	Description	Maintenance
	Mild steel	
25	Hot dip galvanized steel to BS EN 10142 or BS EN 10147, minimum 275g/m² zinc coating weight. PVC/Plastisol coating, 200 microns nominal thickness.	Annual inspection & maintenance in accordance with BS 5427:Part 1, Table 9. Regular cleaning with non-alkaline detergent (polluted/marine areas every 3 months, other areas every 6 months). Renewal of side/end lap seals as necessary.
20	Hot dip aluminium-zinc coated steel to BS EN 10215, minimum 150g/m² zinc coating weight. PVDF/PVF2 coating, 25-50 microns nominal thickness.	Annual inspection & maintenance in accordance with BS 5427:Part 1, Table 9. Regular cleaning with non-alkaline detergent (polluted/marine areas every 3 months,other areas every 6 months). Renewal of side/end lap seals as necessary.
20	Hot dip aluminium-zinc coated steel to BS EN 10215, minimum 150g/m² zinc coating weight. Acrylic/polyester/silicone polyester coating, 25-50 microns nominal thickness.	Annual inspection & maintenance in accordance with BS 5427:Part 1, Table 9. Regular cleaning with non-alkaline detergent (polluted/marine areas every 3 months, other areas every 6 months). Renewal of side/end lap seals as necessary.
15	Hot dip galvanized steel to BS EN 10142 or BS EN 10147, minimum 275g/m² zinc coating weight. PVDF/PVF2 coating, 25-50 microns nominal thickness.	Annual inspection & maintenance in accordance with BS 5427:Part 1, Table 9. Regular cleaning with non-alkaline detergent (polluted/marine areas every 3 months, other areas every 6 months). Renewal of side/end lap seals as necessary.
15	Hot dip galvanized steel to BS EN 10142 or BS EN 10147, minimum 275g/m² zinc coating weight. Acrylic/polyester/silicone polyester coating, 25-50 microns nominal thickness.	Annual inspection & maintenance in accordance with BS 5427:Part 1, Table 9. Regular cleaning with non-alkaline detergent (polluted/marine areas every 3 months, other areas every 6 months). Renewal of side/end lap seals as necessary.
10	Hot dip galvanized steel to BS EN 10142 or BS EN 10147, minimum 350g/m² zinc coating weight.	Annual inspection & maintenance in accordance with BS 5427:Part 1, Table 9. Renewal of side/end lap seals as necessary.
U1	Unclassified, ie mild steel not to relevant BS, or coating/surface protection less than above specifications.	Unclassified.

For Adjustment factors, Assumptions, Notes, Key failure modes and Key durability issues please see overleaf.

2.2a

Profiled metal claddings

Adjustment factors

Galvanized/plated steel fixings: life limited to 25 years.

Assumptions

Coatings may deteriorate during the lives quoted. Recoating is feasible but is not normally practical and is rarely carried out.

Profiled metal cladding may be single skin, double skin with profiled metal liner, insulation and outer skin assembled on site, or factory made composite panels with a rigid foam core.

Design and installation in accordance with BS 5427: Part 1, BS 8200 and in strict accordance with manufacturer's instructions.

Adequate provision for movement to be made in the system when installed.

Fixings to be plated carbon steel or stainless steel to BS 1494:Part 1, or proprietary fixings in accordance with manufacturer's guidance, and to be compatible with other metals used.

Type and spacing of fixing in strict accordance with manufacturer's directions. Heads of plated steel fixings to be protected, eg by push-on caps or integral plastic heads.

Spacing of supports and spacer bars in accordance with manufacturer's load-span tables. Spacer bars, channels, supporting purlins etc to be minimum 275g/m² galvanized steel or no less durable than the cladding.

Provision of vapour control layer, breather membrane and ventilation to built up systems in accordance with the guidance in BS 5427:Part 1.

Sealants to side and end laps (where required by BS 5427:Part 1) to be non-setting gap filling, or pre-formed tape, and compatible with the metal and its coating, and UV resistant if exposed to sunlight.

Provision of movement joints in accordance with BS 5427:Part 1, Table 5.

Coating of reverse side of outer sheet and of liner panels to be appropriate for the environment. Liner panels to be minimum 275g/m² galvanized mild steel, with minimum 20 micron organic coating.

Where two or more metals are used in the cladding construction, they should be compatible or isolated from each other, ie to prevent galvanic corrosion. For further detailed guidance, see BS PD 6484.

Where the roof is designed for regular access, special walkways should be provided to prevent damage to the cladding.

Notes

The life of metal coatings is influenced by the colour of the coating, the orientation of the building, and by the environmental conditions. Detailed guidance on lives and on any guarantee or warranty schemes is available from manufacturers. Proprietary products are available for the re-coating of factory applied coatings.

Corrosion of unprotected cut edges is a common problem which can occur within a relatively short space of time. This process can be delayed by the application of additional edge protection at the time of installation (eg latex system or other products recommended by sheet supplier).

Some fading and chalking of coatings may occur within the life quoted.

Further guidance on the design and installation of profiled metal cladding is available from the Metal Cladding and Roofing Manufacturers' Association and the National Federation of Roofing Contractors.

Key failure modes

Corrosion (particularly at cut edges, fixings and changes in profile), scratching/chalking/blistering/peeling of surface coating, impact/indentation/distortion, elongation of fixing holes, pull-out of fixings, debonding of skin and core (composite panels).

Key durability issues

Base metal, type/thickness of surface protection, maintenance frequency, exposure conditions, colour of coating.

Profiled metal claddings

LOCATIONS - Walls, Roofs

BPG 2 - Walling & Cladding Components

General	Description	Maintenance
Aluminium		
30	Aluminium to BS 1474 (extrusions) or BS EN 485 (fabrications/sheet). Aluminium-zinc alloy anodic coating to both faces. BBA or other 3rd party certified.	Annual inspection & maintenance in accordance with BS 5427:Part 1, Table 9. Renewal of side/end lap seals as necessary.
30	Aluminium to BS 1474 (extrusions) or BS EN 485 (fabrications/sheet). PVDF/PVF2 coating , 25-50 microns nominal thickness.	Annual inspection & maintenance in accordance with BS 5427:Part 1, Table 9. Regular cleaning with non-alkaline detergent (polluted/marine areas every 3 months, other areas every 6 months). Renewal of side/end lap seals as necessary.
30	Aluminium to BS 1474 (extrusions) or BS EN 485 (fabrications/sheet). Acrylic/polyester/silicone polyester coating, 25-50 microns nominal thickness.	Annual inspection & maintenance in accordance with BS 5427:Part 1, Table 9. Regular cleaning with non-alkaline detergent (polluted/marine areas every 3 months, other areas every 6 months). Renewal of side/end lap seals as necessary.
30	Aluminium to BS 1474 (extrusions) or BS EN 485 (fabrications/sheet). Mill finished.	Annual inspection & maintenance in accordance with BS 5427:Part 1, Table 9. Renewal of side/end lap seals as necessary.
U1	Unclassified, ie aluminium not to relevant BS.	Unclassified.

For Adjustment factors, Assumptions, Notes, Key failure modes and Key durability issues please see overleaf.

2.2b

Profiled metal claddings

LOCATIONS - Walls, Roofs

Adjustment factors

Galvanized/plated steel fixings: life limited to 25 years.

Assumptions

Coatings may deteriorate during the lives quoted. Recoating is feasible but is not normally practical and is rarely carried out.

Profiled metal cladding may be single skin, double skin with profiled metal liner, insulation and outer skin assembled on site, or factory made composite panels with a rigid foam core.

Design and installation in accordance with BS 5427: Part 1, BS 8200 and in strict accordance with manufacturer's instructions.

Adequate provision for movement to be made in the system when installed.

Fixings to be plated carbon steel or stainless steel to BS 1494:Part 1, or proprietary fixings in accordance with manufacturer's guidance, and to be compatible with other metals used.

Type and spacing of fixing in strict accordance with manufacturer's directions. Aluminium to be fitted with austenitic stainless steel fixings (which must be isolated from the aluminium in marine or heavily polluted environments due to risk of bimetallic corrosion). Heads of plated steel fixings to be protected, eg by push-on caps or integral plastic heads.

Aluminium cladding to be separated from steel purlins/rails, eg with PVC tape, zinc chromate paint, to prevent bimetallic corrosion.

Spacing of supports and spacer bars in accordance with manufacturer's load-span tables. Spacer bars, channels, supporting purlins etc to be minimum 275g/m² galvanized steel or no less durable than the cladding.

Provision of vapour control layer, breather membrane and ventilation to built up systems in accordance with the guidance in BS 5427:Part 1.

Sealants to side and end laps (where required by BS 5427:Part 1) to be non-setting gap filling, or pre-formed tape, and compatible with the metal and its coating, and UV resistant if exposed to sunlight.

Provision of movement joints in accordance with BS 5427:Part 1, Table 5.

Coating of reverse side of outer sheet and of liner panels to be appropriate for the environment. Liner panels to be minimum 275g/m² galvanized mild steel, with minimum 20 micron organic coating.

In external/damp locations, avoid direct contact between aluminium alloys and timber treated with copper, zinc or mercury based preservatives, Oak, Sweet Chestnut, Douglas Fir, Western Red Cedar, copper alloys (or rainwater run off from), concrete, mortar or soil.

Where two or more metals are used in the cladding construction, they should be compatible or isolated from each other, ie to prevent galvanic corrosion. For further detailed guidance, see BS PD 6484.

Where the roof is designed for regular access, special walkways should be provided to prevent damage to the cladding.

Notes

The life of metal coatings is influenced by the colour of the coating, the orientation of the building, and by the environmental conditions. Detailed guidance on lives and on any guarantee or warranty schemes is available from manufacturers. Proprietary products are available for the re-coating of factory applied coatings.

Corrosion of unprotected cut edges is a common problem which can occur within a relatively short space of time. This process can be delayed by the application of additional edge protection at the time of installation (eg latex system or other products recommended by sheet supplier).

Some fading and chalking of coatings may occur within the life quoted.

Further guidance on the design and installation of profiled metal cladding is available from the Metal Cladding and Roofing Manufacturers' Association and the National Federation of Roofing Contractors.

Key failure modes

Corrosion (particularly at cut edges, fixings and changes in profile), scratching/chalking/blistering/peeling of surface coating, impact/indentation/distortion, elongation of fixing holes, pull-out of fixings, debonding of skin and core (composite panels).

Key durability issues

Base metal, type/thickness of surface protection, maintenance frequency, exposure conditions, colour of coating.

Patent glazing systems

Scope

This section provides data on patent glazing systems for use in non-domestic building types. Patent glazing is defined in BS 5516 as 'a self-draining and ventilated system of dry glazing which does not rely necessarily for its watertightness upon external glazing seals'. It comprises a series of longitudinal supporting members (patent glazing bars), and an infilling of glass or other suitable material.

This section includes aluminium glazing bars and transverse members (from which almost all systems are manufactured), and a range of common glazing materials including double glazed units and plastics. Sealants and gaskets are included in Section 4 of the HAPM Component Life Manual and are therefore excluded from this section.

The following component sub-types are included within this section:

			Page
•	Longitudinal glazing bars:	Aluminium	2.3
•	Transverse supporting members:	Aluminium	2.4
•	Infill materials:	Glass	2.5
		Double glazed units	2.5
		GRP	2.5a
		Polycarbonate	2.5a
		Acrylic	2.5b
		PVC-u	2.5b

Standards cited

BS 952	Glass for glazing.
Part 1:1995	Classification.
BS 1178:1982	Specification for milled lead sheet for building purposes.
BS 1449	Steel plate, sheet and strip.
Part 1:1991	Carbon and carbon-manganese plate, sheet and strip.

BS 1474:1987	Specification for wrought aluminium and aluminium alloys for general engineering purposes; bars, extruded round tubes and sections.
BS 3987:1991(1997)	Specification for anodic oxidation coatings on aluminium for external architectural applications.
BS 4154:various	Corrugated plastic translucent sheets made from thermo-setting polyester resin (glass fibre reinforced).
BS 4842:1984(1991)	Specification for liquid organic coatings application to aluminium alloy extrusions.
BS 4203	Extruded rigid PVC corrugated sheeting.
Part 1:1994	Specification for performance requirements.
BS 5516:1991	Code of practice for design and installation of sloping and vertical patent glazing.
BS 5713:1979(1994)	Specification for hermetically sealed flat double glazing units.
BS 6262:1982	Code of practice for glazing for buildings.
BS 6496:1984(1991)	Specification for powder organic coatings for application and stoving to aluminium alloy extrusions, sheet and preformed sections for external architectural purposes, and for the finish on aluminium alloy extrusions, sheet and preformed sections coated with powder organic coatings.
BS 8118:1991	Structural use of aluminium.
BS PD 6484:1979(1991)	Commentary on corrosion at bimetallic contacts and its alleviation.
BS EN 485: various	Aluminium and aluminium alloys.Sheet, strip and plate.
BS EN 572:1995	Glass in building. Basic soda lime silicate glass products.
BS EN 10143:1993	Continuously hot-dip metal coated steel sheet and strip. Tolerances on dimensions and shape.

Other references/information sources

Patent Glazing Contractors' Association: 'Setting the standard. Patent glazing: notes for the guidance of specifiers'.

Centre for Window & Cladding Technology (CWCT): 'Guide to good practice for facades'.

Patent glazing -
longitudinal glazing bars

LOCATIONS - General

BPG | **2 - Walling & Cladding Components**

General	Description	Maintenance
Aluminium		
30	Extruded aluminium alloy sections to BS 1474. Anodized to BS 3987, minimum 25 micron coating.	Regular cleaning with non-alkaline detergent (polluted/marine areas every 3 months, other areas every 6 months).
25	Extruded aluminium alloy sections to BS 1474. Liquid organic coating to BS 4842, minimum 40 micron thickness.	Regular cleaning with non-alkaline detergent (polluted/marine areas every 3 months, other areas every 6 months). Coatings may require redecoration after 10-15 years.
25	Extruded aluminium alloy sections to BS 1474. Polyester powder coating to BS 6496, minimum 40 micron thickness.	Regular cleaning with non-alkaline detergent (polluted/marine areas every 3 months, other areas every 6 months). Coatings may require redecoration after 10-15 years.
15	Extruded aluminium alloy sections to BS 1474. Mill finished.	Regular cleaning with non-alkaline detergent (polluted/marine areas every 3 months, other areas every 6 months).
U1	Unclassified, ie extruded aluminium alloy sections, not to BS 1474.	Unclassified.

Adjustment factors

Industrial/polluted/marine environment: -5 years.

Assumptions

Extruded aluminium sections to be BS 1474 designation 6063, temper T6 or alternative designation as approved by the manufacturer.

Design and installation in accordance with BS 5516.

Aluminium structural members to comply with BS 8118.

Fasteners to be brass alloy, aluminium, plated/ sherardised mild steel, or austenitic stainless steel, and to comply with the requirements of BS 5516.

Flashings to be one of the following:

– milled lead sheet to BS 1178, minimum thickness 1.8mm (code 4).

– site formed: sheet or strip aluminium to BS 1178, designation 1050A, temper 0, minimum 1.8mm thickness.

– preformed: sheet or strip aluminium to BS EN 485, designation 1200 or alloys designated 3103, 5005 or 5251, minimum 0.9mm thickness.

– steel sheet to BS 1449:Part 1 or galvanized steel sheet to BS EN 10143 and not less than 0.9mm thick.

In external/damp locations, avoid direct contact between aluminium alloys and timber treated with copper, zinc or mercury based preservatives, Oak, Sweet Chestnut, Douglas Fir, Western Red Cedar, copper alloys (or rainwater run off from), concrete, mortar or soil.

Where two or more metals are used in the door construction, they should be compatible, ie to prevent galvanic corrosion. For further detailed guidance, see BS PD 6484.

Notes

The life of metal coatings is influenced by the colour of the coating, the orientation of the building, and by the environmental conditions. Detailed guidance on coating lives and on warranties is generally available from manufacturers.

Key failure modes

Chalking/peeling of surface finish, impact/distortion.

Key durability issues

Type/thickness of surface protection, exposure conditions, maintenance frequency.

TYPE

Patent glazing -
transverse supporting members

LOCATIONS - General

SUB TYPES
Aluminium

Description

Aluminium

General	Description
30	Aluminium alloy sections to BS 1474 (extruded) or BS EN 485 (formed sections). Anodized to BS 3987, minimum 25 micron coating.
25	Aluminium alloy sections to BS 1474 (extruded) or BS EN 485 (formed sections). Liquid organic coating to BS 4842, minimum 40 micron thickness.
25	Aluminium alloy sections to BS 1474 (extruded) or BS EN 485 (formed sections). Polyester powder coating to BS 6496, minimum 40 micron thickness.
15	Aluminium alloy sections to BS 1474 (extruded) or BS EN 485 (formed sections). Mill finished.
U1	Unclassified, ie aluminium alloy sections, not to BS 1474 or BS EN 485.

Maintenance

Regular cleaning with non-alkaline detergent (polluted/marine areas every 3 months, other areas every 6 months).

Regular cleaning with non-alkaline detergent (polluted/marine areas every 3 months, other areas every 6 months). Coatings may require redecoration after 10-15 years.

Regular cleaning with non-alkaline detergent (polluted/marine areas every 3 months, other areas every 6 months). Coatings may require redecoration after 10-15 years.

Regular cleaning with non-alkaline detergent (polluted/marine areas every 3 months, other areas every 6 months).

Unclassified.

Adjustment factors

Industrial/polluted/marine environment: -5 years.

Assumptions

Design and installation in accordance with BS 5516.

Extruded aluminium sections to be BS 1474 designation 6063, temper T6 or alternative designation as approved by the manufacturer.

Fasteners to be brass alloy, aluminium, plated/sherardised mild steel, or austenitic stainless steel, and to comply with the requirements of BS 5516.

Flashings to be one of the following:

- milled lead sheet to BS 1178, minimum thickness 1.8mm (code 4).

- site formed: sheet or strip aluminium to BS 1178, designation 1050A, temper O, minimum 1.8mm thickness.

- preformed: sheet or strip aluminium to BS EN 485, designation 1200 or alloys designated 3103, 5005 or 5251, minimum 0.9mm thickness.

- steel sheet to BS 1449:Part 1 or galvanized steel sheet to BS EN 10143 and not less than 0.9mm thick.

In external/damp locations, avoid direct contact between aluminium alloys and timber treated with copper, zinc or mercury based preservatives, Oak, Sweet Chestnut, Douglas Fir, Western Red Cedar, copper alloys (or rainwater run off from), concrete, mortar or soil.

Where two or more metals are used in the door construction, they should be compatible, ie to prevent galvanic corrosion. For further detailed guidance, see BS PD 6484.

Notes

The life of metal coatings is influenced by the colour of the coating, the orientation of the building, and by the environmental conditions. Detailed guidance on coating lives and on warranties is generally available from manufacturers.

Key failure modes

Chalking/peeling of surface finish, impact/distortion.

Key durability issues

Type/thickness of surface protection, exposure conditions, maintenance frequency.

Patent glazing –

infill materials

LOCATIONS - General

General	Description	Maintenance
	Glass	
35+	Heat soaked thermally toughened laminated glass to BS 952:Part 1and BS EN 572.	Replace glazing gaskets, weatherstripping etc as required.
35+	Heat soaked thermally toughened/thermally toughened body-tinted glass to BS 952:Part 1 and BS EN 572.	Replace glazing gaskets, weatherstripping etc as required.
35+	Heat soaked thermally toughened reflective/thermally toughened opaque glass to BS 952:Part 1 and BS EN 572.	Replace glazing gaskets, weatherstripping etc as required.
35+	Laminated glass to BS 952:Part 1 and BS EN 572.	Replace glazing gaskets, weatherstripping etc as required.
35	Thermally toughened glass to BS 952:Part 1 and BS EN 572.	Replace glazing gaskets, weatherstripping etc as required.
35	Heat strengthened glass to BS 952:Part 1 and BS EN 572.	Replace glazing gaskets, weatherstripping etc as required.
35	Wired glass to BS 952:Part 1 and BS EN 572.	Replace glazing gaskets, weatherstripping etc as required.
35	Annealed glass to BS 952:Part 1 and BS EN 572.	Replace glazing gaskets, weatherstripping etc as required.
U1	Unclassified, ie glass not to BS 952:Part 1 and/or BS EN 572.	Unclassified.
	Double glazed units	
25	Dual seal units Kitemarked to BS 5713.	Replace glazing gaskets, weatherstripping etc as required.
20	Single seal units Kitemarked to BS 5713.	Replace glazing gaskets, weatherstripping etc as required.
15	Dual seal units labelled to BS 5713.	Replace glazing gaskets, weatherstripping etc as required.
15	Single seal units labelled to BS 5713.	Replace glazing gaskets, weatherstripping etc as required.
U1	Unclassified, ie single or dual seal units not labelled to BS 5713.	Unclassified.

For Adjustment factors, Assumptions, Notes, Key failure modes and Key durability issues please see overleaf.

2.5

Patent glazing -
infill materials

LOCATIONS - General

Key failure modes

Fracture, cracking, abrasion, misting of double glazing units (edge seal failure).

Key durability issues

Glazing method, edge seal type and spacer bar width (double glazed units), thickness of glass.

Notes

Insulating infill panels are available which consist of thermally toughened glass with a backing of insulating material, which is either adhered to the glass or retained in a box behind the glass.

To comply with BS 5713 double glazed units must have the BS No. and manufacturer's name or trademark incised on the glass or spacer unit. Kitemarked units must also be marked with the BSI Kitemark emblem.

Detailed guidance on glazing methods for double glazed units is provided in Section 4.2 of the GGF Glazing Manual.

The presence of nickel sulphide inclusions in toughened glass can lead to the risk of 'spontaneous breakage' during normal use. It is widely recognised that heat soaking the glass after toughening, which converts the nickel sulphide to a more stable form, significantly reduces this risk.

For metal infill panels, see curtain walling, p. 2.7.

For glazing/frame sealants and gaskets, see HAPM Component Life Manual p.4.11-4.12a.

Adjustment factors

Double glazed units not gasket glazed (ie units fully bedded): -10 years.

Assumptions

Minimum glass thickness in accordance with BS 6262, with particular regard given to the ability to withstand the calculated design wind pressures and types of location to satisfy safety requirements.

Installation (including glazing method and edge cover) in accordance with BS 6262.

Provision of setting blocks, location blocks and distance pieces in accordance with Glass & Glazing Federation Glazing Manual, figures 7 and 8 (double glazed units).

Patent glazing -
infill materials

LOCATIONS - General

BPG | 2 - Walling & Cladding Components

General	Description	Maintenance
GRP		
30	UV stabilised GRP sheet/section manufactured to BS 4154, minimum weight 2.44kg/m². Outer surface protected with factory applied fluoride-based film.	Periodic cleaning with warm water and mild (non-abrasive) detergent. Inspect fixings (and tighten if necessary) annually. Inspect gaskets annually, replace as necessary.
25	UV stabilised GRP sheet/section manufactured to BS 4154, minimum weight 2.44kg/m² for single skin applications, 1.83kg/m² (ie 1mm thick) for double/triple skin applications. Outer surface protected with factory applied UV protected film.	Periodic cleaning with warm water and mild (non-abrasive) detergent. Application of acrylic or polyester lacquer to restore sheet surface (typically between years 5 and 10, depending upon rate of surface erosion). Inspect fixings (and tighten if necessary) annually. Inspect gaskets annually, replace as necessary.
10	UV stabilised GRP sheet/section manufactured to BS 4154, minimum weight 1.83kg/m² (ie 1mm thick). Outer surface protected with polyester film/gel coat.	Periodic cleaning with warm water and mild (non-abrasive) detergent. Application of acrylic or polyester lacquer to restore sheet surface (typically between years 5 and 10, depending upon rate of surface erosion). Inspect fixings (and tighten if necessary) annually. Inspect gaskets annually, replace as necessary.
U1	Unclassified, ie GRP sheet/section not to BS 4154, and/or less than 1.83kg/m² weight/1mm thickness.	Unclassified.
Polycarbonate		
20	Polycarbonate sheet/section, minimum 1.5mm thick. UV protective surface film/coating to outer face.	Periodic cleaning with warm water and mild (non-abrasive) detergent. Application of acrylic or polyester lacquer to restore sheet surface (typically between years 5 and 10, depending upon rate of surface erosion). Inspect fixings (and tighten if necessary annually. Inspect gaskets annually, replace as necessary.
15	Polycarbonate sheet/section, minimum 1.3mm thick. UV protective surface film/coating to outer face.	Periodic cleaning with warm water and mild (non-abrasive) detergent. Application of acrylic or polyester lacquer to restore sheet surface (typically between years 5 and 10, depending upon rate of surface erosion). Inspect fixings (and tighten if necessary) annually. Inspect gaskets annually, replace as necessary.
10	Polycarbonate sheet/section, less than 1.3mm thick. UV protective surface film/coating to outer face.	Periodic cleaning with warm water and mild (non-abrasive) detergent. Application of acrylic or polyester lacquer to restore sheet surface (typically between years 5 and 10, depending upon rate of surface erosion). Inspect fixings (and tighten if necessary) annually. Inspect gaskets annually, replace as necessary.
U1	Unclassified, ie polycarbonate sheet without UV protective surface film/coating to outer face.	Unclassified.

2.5a | **For Adjustment factors, Assumptions, Notes, Key failure modes and Key durability issues please see overleaf.**

Patent glazing -
infill materials

LOCATIONS - General

BPG 2 - Walling & Cladding Components

Adjustment factors

Use in industrial/polluted/marine environment: -5 years.

Use in contact with Plastisol coated steel, or rainwater wash from Plastisol coating: -5 years (polycarbonate only).

Assumptions

The lives for double/triple skin products are based on the specification of the outer skin. The inner skin(s) may be of a lesser specification (eg not UV stabilised and/or surface protected).

Compatibility of sealant materials to be verified with glazing sheet manufacturer.

Glazing sheet thickness to be appropriate for the application.

Glazing materials must not be painted over with an opaque covering.

Installation (including span, supports and fixing types/ frequency) in accordance with manufacturer's instructions.

Thermoplastic (ie PVC-u, polycarbonate, acrylic) sheet requires oversized fixing holes to accommodate thermal movement.

Notes

Some UV discolouration/yellowing can be expected to occur in most products within 5-10 years (including those with applied surface protection). However, recently introduced fluoride-based films are claimed to offer enhanced resistance to discolouration. GRP containing fire-retardant additives is likely to discolour rapidly.

Polycarbonate is prone to damage (embrittlement, stress cracking) by solvents and by the plasticizer in Plastisol coatings.

For applications where regular foot traffic is expected, a minimum material weight of 5.5kg/m³ is advisable, to ensure safety and resistance to damage (GRP only).

For glazing/frame sealants and gaskets, see HAPM Component Life Manual p.4.11-4.12a.

Key failure modes

UV degradation, weathering/surface erosion, impact, chafing/pull out around fixings, fracture/indentation due to imposed loading.

Key durability issues

UV protection, surface protection, material thickness, adequacy of fixings/supports, provision for thermal movement.

Patent glazing - _____
infill materials

BPG	2 - Walling & Cladding Components	
General	**Description**	**Maintenance**
	Acrylic	
20	Cast/extruded acrylic sheet/section, minimum 3mm thick.	Periodic cleaning with warm water and mild (non-abrasive) detergent. Application of acrylic or polyester lacquer to restore sheet surface (typically between years 5 and 10, depending upon rate of surface erosion). Inspect fixings (and tighten if necessary) annually. Inspect gaskets annually, replace as necessary.
15	Cast/extruded acrylic sheet/section, minimum 2mm thick.	Periodic cleaning with warm water and mild (non-abrasive) detergent. Application of acrylic or polyester lacquer to restore sheet surface (typically between years 5 and 10, depending upon rate of surface erosion). Inspect fixings (and tighten if necessary) annually. Inspect gaskets annually, replace as necessary.
U1	Unclassified, ie cast/extruded acrylic sheet, less than 2mm thick.	Unclassified.
	PVC-u	
10	UV stabilised, extruded/thermoformed PVC-u sheet/section, minimum 1.5mm thick. Extruded sheets/sections to BS 4203:Part 1.	Periodic cleaning with warm water and mild (non-abrasive) detergent. Application of acrylic or polyester lacquer to restore sheet surface (typically between years 5 and 10, depending upon rate of surface erosion). Inspect fixings (and tighten if necessary) annually. Inspect gaskets annually, replace as necessary.
5	UV stabilised, extruded/thermoformed PVC-u sheet/section, minimum 1.3mm thick. Extruded sheets/sections to BS 4203:Part 1.	Periodic cleaning with warm water and mild (non-abrasive) detergent. Application of acrylic or polyester lacquer to restore sheet surface (typically between years 5 and 10, depending upon rate of surface erosion). Inspect fixings (and tighten if necessary) annually. Inspect gaskets annually, replace as necessary.
U1	Unclassified, ie PVC-u sheet/section, not UV stabilised, and/or less than 1.3mm thick, and/or extruded sheets not to BS 4203:Part 1.	Unclassified.

For Adjustment factors, Assumptions, Notes, Key failure modes and Key durability issues please see overleaf.

2.5b

Patent glazing -
infill materials

LOCATIONS - General

2 - Walling & Cladding Components

Adjustment factors

Use in industrial/polluted/marine environment: -5 years.

Factory applied surface protection: +5 years (PVC-u only).

Assumptions

The lives for double/triple skin products are based on the specification of the outer skin. The inner skin(s) may be of a lesser specification (eg not UV stabilised and/or surface protected).

Compatibility of sealant materials to be verified with glazing sheet manufacturer.

Glazing sheet thickness to be appropriate for the application.

Glazing materials must not be painted over with an opaque covering.

Installation (including span, supports and fixing types/frequency) in accordance with manufacturer's instructions.

Thermoplastic (ie PVC-u, polycarbonate, acrylic) sheet requires oversized fixing holes to accommodate thermal movement.

Notes

Some UV discolouration/yellowing can be expected to occur in most products within 5-10 years (including those with applied surface protection). However, recently introduced fluoride-based films are claimed to offer enhanced resistance to discolouration.

For glazing/frame sealants and gaskets, see HAPM Component Life Manual p.4.11-4.12a.

Key failure modes

UV degradation, weathering/surface erosion, impact, chafing/pull out around fixings, fracture/indentation due to imposed loading.

Key durability issues

UV protection, surface protection, material thickness, adequacy of fixings/supports, provision for thermal movement.

Curtain walling system

Continued overleaf

Scope

This section provides data on the most commonly available curtain walling systems for use in non-domestic building types. Curtain walling is defined in the CWCT standard as: 'a form of vertical building enclosure which supports no load other than its own weight and the environmental forces which act upon it'. This might comprise a light carrier framework assembled on site to support infill panels; sections of prefabricated wall fixed directly to the building facade; or prefabricated frames hung onto a site assembled carrier framework.

This section includes metal framing systems, and metal and glazed infill panels. Thermal insulation materials are included in Sections 2 and 3 of the HAPM Component Life Manual. Sealants and gaskets are included in Section 4, and fibre cement and other infill materials are included in Section 2 of the HAPM Manual.

The following component sub-types are included within this section:

		Page
Framing:	Stainless steel	2.6
	Mild steel	2.6
	Aluminium	2.6a
Infill panels:	Stainless steel	2.7
	Mild steel	2.7
	Aluminium	2.7a
	Glass	2.7b
	Double glazed units	2.7b

It should be noted that there is an enormous range of different types and compositions of curtain walling available to UK specifiers. This section aims to cover the most commonly used materials and types.

Standards cited

BS 729:1971(1994)	Specification for hot dip galvanized coatings on iron and steel articles.
BS 952: Part 1:1995	Glass for glazing. Classification.
BS 1449:	Steel plate, sheet and strip.
Part 2:1983	Specification for stainless and heat-resisting steel plate, sheet and strip.
BS 1474:1987	Specification for wrought aluminium and aluminium alloys for general engineering purposes: bars, extruded round tubes and sections.
BS 1706:1990(1996)	Method for specifying electroplated coatings of zinc and cadmium on iron and steel.
BS 2569	Specifications for sprayed metal coatings.
Part 2:1965(1997)	Protection of iron and steel against corrosion and oxidation at elevated temperatures.
BS 3830:1973(1994)	Specification for vitreous enamelled steel building components.
BS 3987:1991(1997)	Specification for anodic oxidation coatings on wrought aluminium for external architectural applications.
BS 4842:1984(1991)	Specification for liquid organic coatings for application to aluminium allot extrusions, sheet and preformed sections for external architectural purposes, and for the finish on aluminium alloy extrusions, sheet and preformed sections coated with liquid organic coatings.
BS 4921:1988 (1994)	Specification for sheradized coatings on iron or steel.
BS 5493:1977	Code of practice for protective coating of iron and steel structures against corrosion.
BS 5713:1979(1994)	Specification for hermetically sealed flat double glazing units.
BS 6262:1982	Code of practice for glazing for buildings.
BS 6496:1984(1991)	Specification for powder organic coatings for application and stoving to aluminium alloy extrusions, sheet and preformed sections for external architectural purposes, and for the finish on aluminium alloy extrusions, sheet and preformed sections coated with powder organic coatings.
BS 6497:1984(1991)	Specification for powder organic coatings for application and stoving to hot-dip galvanized hot-rolled steel sections and preformed steel sheet for windows and associated external architectural purposes, and for the finish on galvanized steel sections and preformed sheet coated with powder organic coatings.

Curtain walling system

Standards cited *continued*

BS 7668:1994	Specification for weldable structural steels. Hot finished structural hollow sections in weather resistant steels.
BS 8118:1991	Structural us of aluminium.
BS 8200:1985	Code of practice for design of non-loadbearing external vertical enclosures of buildings.
BS PD 6484:1979(1991)	Commentary on corrosion at bimetallic contacts and its alleviation.
BS EN 485: various	Aluminium and aluminium alloys. Sheet, strip and plate.
BS EN 572:1995	Glass in building. Basic soda lime silicate glass products.
BS EN 10029:1991	Specification for tolerances on dimensions, shape and mass for hot-rolled steel plates 3mm thick or above.
BS EN 10088-2:1995	Technical delivery conditions for sheet/plate and strip for general purposes.
BS EN 10113:1993	Hot-rolled products in weldable fine grain structural steels.
BS EN 10137:1996	Plates and wide flats made of high yield strength structural steels in the quenched and tempered or precipitation hardened conditions.
BS EN 10142:1991	Specification for continuously hot-dip zinc coated low carbon steel sheet and strip for cold forming; technical delivery conditions.
BS EN 10147:1992	Specification for continuously hot-dip zinc coated structural steel and strip. Technical delivery conditions.
BS EN 10155:1993	Structural steels with improved atmospheric corrosion resistance. Technical delivery conditions.
BS EN 10210:various	Hot finished structural hollow sections of non alloy and fine grain structural steels.
BS EN ISO 3506:various	Mechanical properties of corrosion resistant stainless steel fasteners.

Other references/information sources

Centre for Window & Cladding Technology (CWCT):

- Guide to good practice for facades.
- Standard for curtain walling.
- Test method for curtain walling.

Wilson, M. and Harrison, P. Appraisal and repair of claddings and fixings. Thomas Telford, 1993.

Aluminium Window Association: 'Specification for aluminium alloy curtain walling'.

Curtain walling - framing

BPG	2 - Walling & Cladding Components	
General	**Description**	**Maintenance**
	Stainless steel	
35+	Austenitic stainless steel to BS 1449:Part 2 and BS EN 10088-2, grade 316 (BS EN grade 1.4401).	Clean every 6 months with mild detergent.
30	Austenitic stainless steel to BS 1449:Part 2 and BS EN 10088-2, grade 304 (BS EN grade 1.4301).	Clean every 6 months with mild detergent.
U1	Unclassified, ie stainless steel, not to BS 1449:Part 2/BS EN 10088-2, or less than above specification.	Unclassified.
	Mild steel	
25	Hot rolled sections, hot dip galvanized to BS 729 after all fabrication, minimum zinc coating weight 610g/m² (or appropriate equivalent coating to BS 5493). Polyester powder coated to BS 6497 to a minimum film thickness of 60 microns.	Regular cleaning with non-alkaline detergent (polluted/marine areas every 3 months, other areas every 6 months). Coatings may require redecoration after 10-15 years.
20	Hot rolled/cold formed sections, hot dip galvanized to BS 729 after all fabrication, minimum zinc coating weight 460g/m² (or appropriate equivalent coating to BS 5493). Polyester powder coated to BS 6497 to a minimum film thickness of 60 microns.	Regular cleaning with non-alkaline detergent (polluted/marine areas every 3 months, other areas every 6 months). Coatings may require redecoration after 10-15 years.
20	Cold formed sections, hot dip galvanized before fabrication to BS EN 10142, minimum zinc coating weight 600g/m². Polyester powder coated to BS 6497 to a minimum film thickness of 60 microns.	Regular cleaning with non-alkaline detergent (polluted/marine areas every 3 months, other areas every 6 months). Coatings may require redecoration after 10-15 years.
15	Cold formed sections, hot dip galvanized before fabrication to BS EN 10142, minimum zinc coating weight 450g/m². Polyester powder coated to BS 6497 to a minimum film thickness of 60 microns.	Regular cleaning with non-alkaline detergent (polluted/marine areas every 3 months, other areas every 6 months). Coatings may require redecoration after 10-15 years.
10	Cold formed sections, hot dip galvanized before fabrication to BS EN 10142, minimum zinc coating weight 275g/m². Polyester powder coated to BS 6497 to a minimum film thickness of 60 microns.	Regular cleaning with non-alkaline detergent (polluted/marine areas every 3 months, other areas every 6 months). Coatings may require redecoration after 10-15 years.
U1	Unclassified, ie mild steel, surface protection less than above specifications.	Unclassified.

For Adjustment factors, Assumptions, Notes, Key failure modes and Key durability issues please see overleaf.

2.6

TYPE

Curtain walling - framing

LOCATIONS - General

SUB TYPES
Stainless steel
Mild steel
Aluminium

Adjustment factors

Industrial/polluted/marine environment: -5 years (except 316 grade austenitic stainless steel).

Assumptions

Design in accordance with Centre for Window and Cladding Technology (CWCT) Standard for curtain walling and with BS 8200.

Mild steel framing sections/reinforcement to BS 7668, BS EN 10029, BS EN 10113, BS EN 10137, BS EN 10155 or BS EN 10210, in a thickness suitable for the application.

Fixing bolts, anchors, brackets, screws to be manufactured from one of the following:

- stainless steel, grade A2 (grade 304 for general use) or grade A4 (grade 316 for severely corrosive environments) of BS EN ISO 3506;

- mild steel, hot dip galvanized to BS 729, sherardised to BS 4921 Class 1 with a passivation post-treatment, zinc plated to BS 1706 classification Zn12, sprayed zinc to BS 569:Part 2 class Zn7 (note: these finishes are not strictly comparable).

- brackets, rivets, shear pins etc can also be provided from aluminium alloy to BS 1474.

With galvanized hollow sections it is essential to provide holes for venting and drainage, and to ensure that the internal surfaces are fully coated with zinc (steel only).

For coated metals, the lives quoted represent the lives to first maintenance. Whether maintenance is possible or even probable will depend on the feasibility of access to ALL surfaces. In most cases such access will not be possible.

Where two or more metals are used in the wall construction, they should be compatible, ie to prevent galvanic corrosion. For further detailed guidance, see BS PD 6484.

Notes

The life of metal coatings is influenced by the colour of the coating, the orientation of the building, and by the environmental conditions. Detailed guidance on lives and on guarantees and warranties is likely to be available from manufacturers.

Further guidance on the specification and testing of curtain walling is provide in the CWCT publications: 'Guide to good practice for facades' and 'Test methods for curtain walling'.

Key failure modes

Corrosion, scratching/chalking/peeling of surface finish, impact/indentation/distortion, fixing failure.

Key durability issues

Base metal, type/thickness of surface protection, maintenance frequency, exposure conditions.

Curtain walling - framing

LOCATIONS - General

2 - Walling & Cladding Components

General	Description	Maintenance
	Aluminium	
30	Extruded aluminium alloy sections to BS 1474. Anodized to BS 3987, minimum 25 micron coating.	Regular cleaning with non-alkaline detergent (polluted/marine areas every 3 months, other areas every 6 months).
25	Extruded aluminium alloy sections to BS 1474. Liquid organic coating to BS 4842 after all fabrication, minimum 40 micron thickness.	Regular cleaning with non-alkaline detergent (polluted/marine areas every 3 months, other areas every 6 months). Coatings may require redecoration after 10-15 years.
25	Extruded aluminium alloy sections to BS 1474. Polyester powder coating to BS 6496 after all fabrication, minimum 40 micron thickness.	Regular cleaning with non-alkaline detergent (polluted/marine areas every 3 months, other areas every 6 months). Coatings may require redecoration after 10-15 years.
15	Extruded aluminium alloy sections to BS 1474. Mill finished.	Regular cleaning with non-alkaline detergent (polluted/marine areas every 3 months, other areas every 6 months).
U1	Unclassified, ie aluminium alloy sections, not to BS 1474.	Unclassified.

For Adjustment factors, Assumptions, Notes, Key failure modes and Key durability issues please see overleaf.

2.6a

Curtain walling - framing

LOCATIONS - General

Adjustment factors

Industrial/polluted/marine environment: -5 years.

Assumptions

Design in accordance with Centre for Window and Cladding Technology (CWCT) Standard for curtain walling and with BS 8200.

In external/damp locations, avoid direct contact between aluminium alloys and timber treated with copper, zinc or mercury based preservatives, Oak, Sweet Chestnut, Douglas Fir, Western Red Cedar, copper alloys (or rainwater run off from), concrete, mortar or soil.

Aluminium structural members to comply with BS 8118.

Fixing bolts, anchors, brackets, screws to be manufactured from one of the following:

- stainless steel, grade A2 (grade 304 for general use) or grade A4 (grade 316 for severely corrosive environments) of BS EN ISO 3506;

- mild steel, hot dip galvanized to BS 729, sherardised to BS 4921 Class 1 with a passivation post-treatment, zinc plated to BS 1706 classification Zn7 (note: these finishes are not strictly comparable).

- brackets, rivets, shear pins etc can also be provided from aluminium alloy to BS 1474.

For coated metals, the lives quoted represent the lives to first maintenance. Whether maintenance is possible or even probable will depend on the feasibility of access to ALL surfaces. In most cases such access will not be possible.

Where two or more metals are used in the wall construction, they should be compatible, ie to prevent galvanic corrosion. For further detailed guidance, see BS PD 6484.

Notes

The life of metal coatings is influenced by the colour of the coating, the orientation of the building, and by the environmental conditions. Detailed guidance on lives and on guarantees and warranties is likely to be available from manufacturers.

Further guidance on the specification and testing of curtain walling is provide in the CWCT publications: 'Guide to good practice for facades' and 'Test methods for curtain walling'.

Key failure modes

Corrosion, scratching/chalking/peeling of surface finish, impact/indentation/distortion, fixing failure.

Key durability issues

Base metal, type/thickness of surface protection, maintenance frequency, exposure conditions.

Curtain walling - _____
infill panels

LOCATIONS - General

BPG 2 - Walling & Cladding Components

General	Description	Maintenance
Stainless steel		
35+	Austenitic stainless steel to BS 1449:Part 2 and BS EN 100088-2, grade 316 (BS EN grade 1.4401).	Clean every 6 months with mild detergent. Renewal of sealants/gaskets as necessary.
30	Austenitic stainless steel to BS 1449:Part 2 and BS EN 100088-2, grade 304 (BS EN grade 1.4301). Not suitable for use in exposed marine or aggressive environment.	Clean every 6 months with mild detergent. Renewal of sealants/gaskets as necessary.
U1	Unclassified, ie stainless steel, not to BS 1449:Part 2/BS EN 100088-2, or less than above specification.	Unclassified.
Mild steel		
30	Vitreous enamelled steel to BS 3830, classified not lower than Class B when tested to BS 1344:Part 2.	Regular cleaning with non-alkaline detergent (polluted/marine areas every 3 months, other areas every 6 months). Renewal of sealants/gaskets as necessary.
25	Pre-galvanized mild steel to BS EN 10142, minimum zinc coating weight 610g/m². Factory applied organic coating, 25-50 micron nominal thickness.	Regular cleaning with non-alkaline detergent (polluted/marine areas every 3 months, other areas every 6 months). Renewal of sealants/gaskets as necessary.
25	Hot dip galvanized steel to BS EN 10142 (pre-galvanized) or BS EN 10147 (post-galvanized), minimum zinc coating weight 275g/m². PVC/plastisol coating, 200 micron nominal thickness.	Regular cleaning with non-alkaline detergent (polluted/marine areas every 3 months, other areas every 6 months). Renewal of sealants/gaskets as necessary.
20	Hot dip galvanized steel to BS EN 10142 (pre-galvanized) or BS EN 10147 (post- galvanized), minimum zinc coating weight 450g/m². Factory applied organic coating, 25- 50 micron nominal thickness.	Regular cleaning with non-alkaline detergent (polluted/marine areas every 3 months, other areas every 6 months). Renewal of sealants/gaskets as necessary.
15	Hot dip galvanized steel to BS EN 10142 (pre-galvanized) or BS EN 10147 (post-galvanized), minimum zinc coating weight 275g/m². Factory applied organic coating, 25-50 micron nominal thickness.	Regular cleaning with non-alkaline detergent (polluted/marine areas every 3 months, other areas every 6 months). Renewal of sealants/gaskets as necessary.
U1	Unclassified, ie mild steel, surface protection less than above specifications.	Unclassified.

For Adjustment factors, Assumptions, Notes, Key failure modes and Key durability issues please see overleaf.

Curtain walling - infill panels

LOCATIONS - General

Adjustment factors

Industrial/polluted/marine environment: -5 years (except 316 grade austenitic stainless steel).

Assumptions

Design in accordance with Centre for Window and Cladding Technology (CWCT) Standard for curtain walling and with BS 8200.

Fixing bolts, anchors, brackets, screws to be manufactured from one of the following:

– stainless steel, grade A2 (grade 304 for general use) or grade A4 (grade 316 for severely corrosive environments) of BS EN ISO 3506;

– mild steel, hot dip galvanized to BS 729, sherardised to BS 4921 Class 1 with a passivation post-treatment, zinc plated to BS 1706 classification Zn12, sprayed zinc to BS 2569:Part 2 class Zn7 (note: these finishes are not strictly comparable).

– brackets, rivets, shear pins etc can also be provided from aluminium alloy to BS 1474.

Coatings may deteriorate during the lives quoted. Recoating is feasible but is not normally practical and is rarely carried out.

Infill panels may be single skin, double skin with metal liner, insulation and outer skin assembled on site, or factory made composite panels with a rigid foam core.

Where two or more metals are used in the wall construction, they should be compatible, ie to prevent galvanic corrosion. For further detailed guidance, see BS PD 6484.

Notes

The life of metal coatings is influenced by the colour of the coating, the orientation of the building, and by the environmental conditions. Detailed guidance on lives and on any guarantee or warranty schemes is available from manufacturers. Proprietary products are available for the re-coating of factory applied coatings.

Further guidance on the specification and testing of curtain walling is provide in the CWCT publications: 'Guide to good practice for facades' and 'Test methods for curtain walling'.

For sealants and gaskets, see HAPM Component Life Manual p.4.11-4.12a.

Key failure modes

Corrosion, scratching/chalking/peeling of surface finish, impact/ indentation/distortion.

Key durability issues

Base metal, type/thickness of surface protection, maintenance frequency, exposure conditions.

Curtain walling -
infill panels

LOCATIONS - General

BPG	2 - Walling & Cladding Components		
General	**Description**		**Maintenance**
	Aluminium		
30	Aluminium alloy to BS EN 485. Anodized to BS 3987, minimum 25 micron coating.		Regular cleaning with non-alkaline detergent (polluted/marine areas every 3 months, other areas every 6 months). Renewal of sealants/gaskets as necessary.
25	Aluminium alloy to BS EN 485. Liquid organic coating to BS 4842 after all fabrication, minimum 40 micron thickness.		Regular cleaning with non-alkaline detergent (polluted/marine areas every 3 months, other areas every 6 months). Renewal of sealants/gaskets as necessary.
25	Aluminium alloy to BS EN 485. Polyester powder coating to BS 6496 after all fabrication, minimum 40 micron thickness.		Regular cleaning with non-alkaline detergent (polluted/marine areas every 3 months, other areas every 6 months). Renewal of sealants/gaskets as necessary.
15	Aluminium alloy to BS EN 485. Mill finished.		Regular cleaning with non-alkaline detergent (polluted/marine areas every 3 months, other areas every 6 months). Renewal of sealants/gaskets as necessary.
U1	Unclassified, ie aluminium alloy, not to BS EN 485.		Unclassified.

For Adjustment factors, Assumptions, Notes, Key failure modes and Key durability issues please see overleaf.

2.7a
Aluminium

Curtain walling -
infill panels

LOCATIONS - General

Adjustment factors

Industrial/polluted/marine environment: -5 years.

Assumptions

Design in accordance with Centre for Window and Cladding Technology (CWCT) Standard for curtain walling and with BS 8200.

Fixing bolts, anchors, brackets, screws to be manufactured from one of the following:

– stainless steel, grade A2 (grade 304 for general use) or grade A4 (grade 316 for severely corrosive environments) of BS EN ISO 3506;

– mild steel, hot dip galvanized to BS 729, sherardised to BS 4921 Class 1 with a passivation post-treatment, zinc plated to BS 1706 classification Zn12, sprayed zinc to BS 2569:Part 2 class Zn7 (note: these finishes are not strictly comparable).

– brackets, rivets, shear pins etc can also be provided from aluminium alloy to BS 1474.

Coatings may deteriorate during the lives quoted. Recoating is feasible but is not normally practical and is rarely carried out.

Infill panels may be single skin, double skin with metal liner, insulation and outer skin assembled on site, or factory made composite panels with a rigid foam core.

In external/damp locations, avoid direct contact between aluminium alloys and timber treated with copper, zinc or mercury based preservatives, Oak, Sweet Chestnut, Douglas Fir, Western Red Cedar, copper alloys (or rainwater run off from), concrete, mortar or soil.

Where two or more metals are used in the wall construction, they should be compatible, ie to prevent galvanic corrosion. For further detailed guidance, see BS PD 6484.

Notes

The life of metal coatings is influenced by the colour of the coating, the orientation of the building, and by the environmental conditions. Detailed guidance on lives and on any guarantee or warranty schemes is available from manufacturers. Proprietary products are available for the re-coating of factory applied coatings.

Further guidance on the specification and testing of curtain walling is provide in the CWCT publications: 'Guide to good practice for facades' and 'Test methods for curtain walling'.

For sealants and gaskets, see HAPM Component Life Manual p.4.11-4.12a.

Key failure modes

Corrosion, scratching/chalking/peeling of surface finish, impact/ indentation/distortion.

Key durability issues

Base metal, type/thickness of surface protection, maintenance frequency, exposure conditions.

Curtain walling -
infill panels

LOCATIONS - General

BPG 2 - Walling & Cladding Components

General	Description	Maintenance
	Glass	
35+	Heat soaked thermally toughened laminated glass to BS 952:Part 1 and BS EN 572.	Replace glazing gaskets, weatherstripping etc as required.
35+	Heat soaked thermally toughened/thermally toughened body-tinted glass to BS 952:Part 1 and BS EN 572.	Replace glazing gaskets, weatherstripping etc as required.
35+	Heat soaked thermally toughened reflective/thermally toughened opaque glass to BS 952:Part 1 and BS EN 572.	Replace glazing gaskets, weatherstripping etc as required.
35+	Laminated glass to BS 952:Part 1 and BS EN 572.	Replace glazing gaskets, weatherstripping etc as required.
35	Thermally toughened glass to BS 952:Part 1 and BS EN 572.	Replace glazing gaskets, weatherstripping etc as required.
35	Heat strengthened glass to BS 952:Part 1 and BS EN 572.	Replace glazing gaskets, weatherstripping etc as required.
35	Annealed glass to BS 952:Part 1 and BS EN 572.	Replace glazing gaskets, weatherstripping etc as required.
U1	Unclassified, ie glass not to BS 952:Part 1 and/or BS EN 572.	Unclassified.
	Double glazed units	
25	Dual seal units Kitemarked to BS 5713.	Replace glazing gaskets, weatherstripping etc as required.
20	Single seal units Kitemarked to BS 5713.	Replace glazing gaskets, weatherstripping etc as required.
15	Dual seal units labelled to BS 5713.	Replace glazing gaskets, weatherstripping etc as required.
15	Single seal units labelled to BS 5713.	Replace glazing gaskets, weatherstripping etc as required.
U1	Unclassified, ie single or dual seal units not labelled to BS 5713.	Unclassified.

For Adjustment factors, Assumptions, Notes, Key failure modes and Key durability issues please see overleaf.

2.7b

Curtain walling -
infill panels

LOCATIONS - General

Key failure modes

Fracture, cracking, abrasion, misting of double glazing units (edge seal failure).

Key durability issues

Glazing method, edge seal type and spacer bar width (double glazed units), thickness of glass.

Notes

Insulating infill panels are available which consist of thermally toughened glass with a backing of insulating material, which is either adhered to the glass or retained in a box behind the glass.

To comply with BS 5713 double glazed units must have the BS No. and manufacturer's name or trademark incised on the glass or spacer unit. Kitemarked units must also be marked with the BSI Kitemark emblem.

Detailed guidance on glazing methods for double glazed units is provided in Section 4.2 of the GGF Glazing Manual.

The presence of nickel sulphide inclusions in toughened glass can lead to the risk of 'spontaneous breakage' during normal use. It is widely recognised that heat soaking the glass after toughening, which converts the nickel sulphide to a more stable form, significantly reduces this risk.

Adjustment factors

Double glazed units not gasket glazed/fully drained (ie units fully bedded): -10 years.

Assumptions

Minimum glass thickness in accordance with BS 6262, with particular regard given to the ability to withstand the calculated design wind pressures and types of location to satisfy safety requirements.

Installation (including glazing method and edge cover) in accordance with BS 6262.

Provision of setting blocks, location blocks and distance pieces in accordance with Glass & Glazing Federation Glazing Manual, figures 7 and 8 (double glazed units).

Internal partitions

Scope

This section provides data on common types of internal partitions systems for non-domestic building types. It includes fixed and demountable partition systems and also rigid panels forming part of a folding or sliding partition system. Low level, movable screens are not included in this section.

The range of different partition systems currently available to specifiers is considerable and this section attempts to cover only the most common framing and facing/infill materials. Whilst in practice the specifier will often select a complete partitioning system from a single supplier, for the purposes of this manual separate sections are provided for framing materials and infill panels/facing and lining materials. This separation enables coverage of a broader range of systems than would otherwise be possible.

The following component sub-types are included within this section:

		Page
Concealed metal studwork		2.8
Metal framing:	Aluminium	2.9
	Steel	2.9
Metal panels:	Stainless steel	2.10
	Mild steel	2.10
	Aluminium	2.10
Timber based panels:	Laminate faced	2.11
	Timber veneer/fabric/ vinyl faced	2.11

Gypsum based plasterboards, gypsum fibreboard and calcium silicate board are included in Section 2 of the HAPM Component Life Manual. Timber studwork is to be included in a future update of the HAPM Manual.

Standards cited

BS 729: 1971 (1994) Specification for hot-dip galvanised coatings on iron and steel articles

BS 970 Specification for wrought steels for mechanical and allied engineering purposes

Part 1: 1996 General inspection and testing procedures and specific requirements for carbon, carbon manganese, alloy and stainless steels

BS 1230 Gypsum plasterboard

Part 1: 1985 (1994) Specification for plasterboard excluding materials submitted to secondary operations

BS 1449:

Part 1: 1991 Steel plate, sheet and strip

Part 2: 1983 Carbon and carbon-manganese plate, sheet and strip

BS 1474: 1987 Specification for stainless and heat resisting steel plate, sheet and strip

Specification for wrought aluminium and aluminium alloy for general engineering purposes : bars, extruded round tubes and sections

BS 1615: 1987 (1994) Method for specifying anodic oxidation coatings on aluminium and its alloys

BS 2994: 1976 (1987) Specification for cold rolled steel sections

BS 4842: 1984 (1991) Specification for liquid organic coatings for application to aluminium alloy extrusions, sheet and preformed sections for external architectural purposes, and for the finish on aluminium alloy extrusions, sheet and preformed sections coated with liquid organic coatings

BS 4965: 1991 (1996) Specification for decorative laminated plastic sheet veneered boards and panels

BS 5234 Partitions (including matching linings)

Part 1: 1992 Code of Practice for design and installation

Part 2: 1992 Specification for performance requirements for strength and robustness including methods of test

BS 6496: 1984 (1991) Specification for powder organic coatings for application and stoving to aluminium alloy extrusions, sheet and preformed sections for external architectural purposes, and for the finish on aluminium alloy extrusions, sheet and preformed sections coated with powder organic coatings

Continued overleaf

Internal partitions

Standards cited *continued*

BS 6566:	Plywood
Part 8: 1985 (1991)	Specification for bond performance of veneer plywood
BS 7332: 1990	Specification for decorative continuous laminates (DCL) based on thermosetting resins
BS 7364: 1990	Specification for galvanised steel studs and channels for stud and stud sheet partitions and linings using screw fixed gypsum wallboards
BS EN 438	Decorative high pressure laminates (HPL) sheets based on thermosetting resins
Part 1: 1991 (1996)	Specifications
BS EN 485: 1994/5	Aluminium and aluminium alloy. Sheet, strip and plate
BS EN 622: 1987	Fibreboards - specifications
BS EN 633: 1994	Cement bonded particleboards - definition and classification
BS EN 634: 1994	Cement bonded particleboards - specification
BS EN 636: 1997	Plywood. Specifications.
BS EN 10142: 1991	Specification for continuously hot-dip zinc coated low carbon steel sheet and strip for cold forming: technical delivery conditions
BS EN 10147: 1992	Specification for continuously hot-dip zinc coated structural steel sheet and strip. Technical delivery conditions.

Other references/information sources

HAPM Component Life Manual pages 2.30 - 2.31 (plasterboards).

Internal partitions -
concealed metal studwork

2 - Walling & Cladding Components

General	Description	Maintenance
	Concealed metal studwork	
35+	Galvanized steel studs/channels to BS 7364 for use in gypsum wallboard faced partitions. Pre-galvanized steel to BS EN 10142, minimum 275g/m² zinc coating weight.	None.
35	Pre-galvanized steel studs/channels to BS EN 10142, minimum 275g/m² zinc coating weight.	None.
U1	Unclassified, ie pre-galvanized steel studs/channels not to BS EN 10142, and/or less than 275g/m² zinc coating weight, galvanized steel studs/channels not to BS 7364, for use in gypsum wallboard faced partitions.	Unclassified.

Adjustment factors

None.

Assumptions

Design and installation of partition system in accordance with BS 5234:Part 1.

Design for appropriate use category in accordance with BS 5234:Part 2. Partitions for use in office locations to be minimum 'medium duty'. Partitions for use in industrial or public circulation areas to be minimum 'heavy duty'.

Gypsum wallboard to BS 1230:Part 1.

Where partitions are built on platform floors and/or rely on suspended ceilings for their stability, the suitability for

loading of these elements should be checked. The fixing of a partition to the underside of a suspended ceiling may require additional fixing supports provided above the suspended ceiling. On platform floors, partitions should be located at the positions recommended by the floor manufacturer and within the maximum recommended loadings. Otherwise, additional supports may be required.

Fixings to be at least galvanized steel or non ferrous and compatible with other metals used.

Surfaces likely to encounter moisture from splashes (eg around sinks) should be protected by tiling or other suitable material.

Notes

BS 7364 applies only to steel studwork for use in partitions faced in gypsum wallboard.

BS 5234:Part 2 defines four use categories for internal partitions (light, medium, heavy, severe duty), along with strength requirements and tests for each category. Tests include stiffness, impact, and stability under loading.

For gypsum based plasterboard, gypsum fibreboard, calcium silicate board, see HAPM Component Life Manual p. 2.30-2.31.

Key failure modes

Impact, overloading, instability, loss/pull out of fixings, corrosion.

Key durability issues

Design for intended use (ie BS 5234 strength class), adequacy of support/fixings, protection against moisture.

Internal partitions –
metal framing

LOCATIONS - General

BPG

2 - Walling & Cladding Components

General	Description	Maintenance
Aluminium		
35+	Aluminium to BS 1474 (extrusions) or BS EN 485 (fabrications/sheet). Minimum 25 micron liquid organic coating to BS 4842 or powder coating to BS 6496.	None.
35+	Aluminium to BS 1474 (extrusions) or BS EN 485 (fabrications/sheet). Anodized to BS 1615, minimum 15 micron coating.	None.
35	Aluminium to BS 1474 (extrusions) or BS EN 485 (fabrications/sheet). Anodized to BS 1615, minimum 10 micron coating.	None.
30	Aluminium to BS 1474 (extrusions) or BS EN 485 (fabrications/sheet). Anodized to BS 1615, minimum 5 micron coating.	None.
25	Aluminium to BS 1474 (extrusions) or BS EN 485 (fabrications/sheet). Mill finished.	None.
U1	Unclassified, ie aluminium, not to relevant BS.	Unclassified.
Steel		
35	Post-galvanized steel to BS EN 10147, or BS 2994, or to BS 1449:Part 1 and galvanized to BS 729. Minimum 450g/m² zinc coating weight.	None.
35	Pre-galvanized steel to BS EN 10142 or to BS 1449:Part 1 and galvanized to BS 729. Minimum 450g/m² zinc coating weight and factory applied organic/stoved enamel coating.	None.
25	Post-galvanized steel to BS EN 10147, or BS 2994, or to BS 1449:Part 1 and galvanized to BS 729. Minimum 275g/m² zinc coating weight.	None.
25	Pre-galvanized steel to BS EN 10142, or to BS 1449:Part 1 and galvanized to BS 729. Minimum 275g/m² zinc coating weight and factory applied organic/stoved enamel coating.	None.
15	Pre-galvanized steel to BS EN 10142, or to BS 1449:Part 1 and galvanized to BS 729. Minimum 275g/m² zinc coating weight.	None.
U1	Unclassified, ie galvanized steel not to relevant BS and/or zinc coating less than 275g/m².	Unclassified.

For Adjustment factors, Assumptions, Notes, Key failure modes and Key durability issues please see overleaf.

2.9

Internal partitions –
metal framing

2 - Walling & Cladding Components

Adjustment factors

Framing fully enclosed behind panel/sheet: +5 years.

Exposed framing used in damp/humid environments, eg kitchens, bathrooms: -5 years.

Life limited to 10 years if plain steel fixings used.

Assumptions

Single skin panels to be stiffened at regular intervals with metal stiffeners/channels.

Design and installation of partition system in accordance with BS 5234:Part 1.

Design for appropriate use category in accordance with BS 5234:Part 2. Partitions for use in office locations to be minimum 'medium duty'. Partitions for use in industrial or public circulation areas to be minimum heavy duty.

Where partitions are built on platform floors and/or rely on suspended ceilings for their stability, the strength of these elements should be checked. The fixing of a partition to the underside of a suspended ceiling may require additional fixing supports provided above the suspended ceiling. On platform floors, partitions should be located at the positions recommended by the floor manufacturer and within the maximum recommended loadings. Otherwise, additional supports may be required.

Fixings to be at least galvanized steel or non ferrous and compatible with other metals used.

Surfaces likely to encounter moisture from splashes (eg around sinks) should be protected by tiling or other suitable material.

Notes

Metal framing may be concealed behind the panels/sheeting of the partition, or exposed, with infill panels between.

BS 5234:Part 2 defines four use categories for internal partitions (light, medium, heavy, severe duty), along with strength requirements and tests for each category. Tests include hard and soft body impact, stiffness and stability under loading.

Where a partition is required to support additional loads (such as lightweight fittings, wash basins, wall cupboards), BS 5234:Part 1 recommends that a sample of the partition is first tested using the methods in BS 5234:Part 2.

The range of options for protecting metals is huge, therefore the options listed above are indicative only.

For gypsum based plasterboard, gypsum fibreboard, calcium silicate board, see HAPM Component Life Manual p. 2.30-2.31.

Key failure modes

Impact, overloading, instability, corrosion, chalking/peeling/abrasion of decorative coating, loss/pull out of fixings.

Key durability issues

Design for intended use (ie BS 5234 strength class), adequacy of support/fixings, protection against moisture, type/thickness of metal coating.

2.9

Internal partitions –
metal panels

BPG 2 - Walling & Cladding Components

General	Description	Maintenance
	Stainless steel	
25	Proprietary panels, austenitic stainless steel to BS 1449:Part 2 or BS 970:Part 1, grades 316 or 304.	Regular cleaning with mild detergent.
20	Proprietary panels, ferritic stainless steel to BS 1449:Part 2 or BS 970:Part 1, grade 430.	Regular cleaning with mild detergent.
U1	Unclassified, ie proprietary panels, stainless steel, not to BS 1449:Part 2 or BS 970:Part 1.	Unclassified.
	Mild steel	
25	Proprietary panels, post-galvanized steel to BS EN 10147 or BS 2994, or to BS 1449:Part 1 and galvanized to BS 729. Minimum 450g/m² zinc coating weight.	Regular cleaning with mild detergent.
25	Proprietary panels, pre-galvanized steel to BS EN 10142 or to BS 1449:Part 1 and galvanized to BS 729. Minimum 450g/m² zinc coating weight and factory applied organic/stoved enamel coating.	Regular cleaning with mild detergent.
15	Proprietary panels, post-galvanized steel to BS EN 10147 or BS 2994, or to BS 1449:Part 1 and galvanized to BS 729. Minimum 275g/m² zinc coating weight.	Regular cleaning with mild detergent.
15	Proprietary panels, pre-galvanized steel to BS EN 10142 or to BS 1449:Part 1 and galvanized to BS 729. Minimum 275g/m² zinc coating weight and factory applied organic/stoved enamel coating.	Regular cleaning with mild detergent.
5	Proprietary panels, pre-galvanized steel to BS EN 10142 or to BS 1449:Part 1 and galvanized to BS 729. Minimum 275g/m² zinc coating weight.	Regular cleaning with mild detergent.
U1	Unclassified, ie proprietary panels, galvanized steel not to relevant BS and/or zinc coating less than 275g/m².	Unclassified.
	Aluminium	
25	Proprietary panels, aluminium to BS 1474 (extrusions) or BS EN 485 (fabrications/sheet). Minimum 25 micronliquid organic coating to BS 4842 or powder coating to BS 6496.	Regular cleaning with mild detergent.
25	Proprietary panels, aluminium to BS 1474 (extrusions) or BS EN 485 (fabrications/sheet). Anodized to BS 1615, minimum 15 micron coating.	Regular cleaning with mild detergent.
20	Proprietary panels, aluminium to BS 1474 (extrusions) or BS EN 485 (fabrications/sheet). Anodized to BS 1615, minimum 10 micron coating.	Regular cleaning with mild detergent.
15	Proprietary panels, aluminium to BS 1474 (extrusions) or BS EN 485 (fabrications/sheet). Anodized to BS 1615, minimum 5 micron coating.	Regular cleaning with mild detergent.
10	Proprietary panels, aluminium to BS 1474 (extrusions) or BS EN 485 (fabrications/sheet). Mill finished.	Regular cleaning with mild detergent.
U1	Unclassified, ie proprietary panels, aluminium, not to relevant BS.	Unclassified.

For Adjustment factors, Assumptions, Notes, Key failure modes and Key durability issues please see overleaf.

2.10

Internal partitions –
metal panels

Adjustment factors

Partition system Kitemarked to BS 5234:Part 1: +5 years.

Metal panels continuously supported, (ie backing of particleboard, plasterboard, plywood, rigid insulation etc): +5 years.

Life limited to 10 years if plain steel fixings used.

Use in damp/humid environments, eg kitchens, bathrooms: -5 years.

Panels used as part of folding/sliding partition system: -5 years.

Assumptions

Single skin panels to be stiffened at regular intervals with metal channels.

Design and installation of partition system in accordance with BS 5234:Part 1.

Design for appropriate use category in accordance with BS 5234:Part 2. Partitions for use in office locations to be minimum 'medium duty'. Partitions for use in industrial or public circulation areas to be minimum heavy duty.

All concealed metal channels/framing members to be minimum galvanized steel with 275g/m² zinc coating weight, or mill finished/anodized aluminium. Exposed metal framing members should have a life assessment at least equivalent to the metal panels (see P.2.2 (framing) above).

Where partitions are built on platform floors and/or rely on suspended ceilings for their stability, the strength of these elements should be checked. The fixing of a partition to the underside of a suspended ceiling may require additional fixing supports provided above the suspended ceiling. On platform floors, partitions should be located at the positions recommended by the floor manufacturer and within the maximum recommended loadings. Otherwise, additional supports may be required.

Fixings to be at least galvanized steel or non ferrous and compatible with other metals used.

Surfaces likely to encounter moisture from splashes (eg around sinks) should be protected by tiling or other suitable material.

Notes

BS 5234:Part 2 defines four use categories for internal partitions (light, medium, heavy, severe duty), along with strength requirements and tests for each category. Tests include hard and soft body impact, stiffness and stability under loading.

Where a partition is required to support additional loads (such as lightweight fittings, wash basins, wall cupboards), BS 5234:Part 1 recommends that a sample of the partition is first tested using the methods in BS 5234:Part 2.

The range of options for protecting steel is huge, therefore the options listed above are indicative only.

For gypsum based plasterboard, gypsum fibreboard, calcium silicate board, see HAPM Component Life Manual p. 2.30-2.31.

Key failure modes

Impact/indentation, loss/pull out of fixings, corrosion, chalking/peeling/abrasion of decorative coating.

Key durability issues

Design for intended use (ie BS 5234 strength class), adequacy of support/fixings/stiffening, protection against moisture, type/thickness of metal coating.

Internal partitions –
timber based panels

LOCATIONS - General

SUB TYPES
Laminate faced panels
Veneer/fabric/vinyl faced panels

BPG | 2 - Walling & Cladding Components

General	Description	Maintenance
Laminate faced panels		
30	High pressure laminate faced boards/panels to BS 4965 durability class D4 (ie plywood type WBP to BS 6566:Part 8, Portland cement bonded particleboard to BS EN 633, non-asbestos mineral board), surface laminate to BS EN 438:Part 1, HG (high performance grade), or BS 7332 grade G.	Regular cleaning with mild detergent.
25	High pressure laminate faced boards/panels to BS 4965. Core material to BS 4965 durability class D4 (ie plywood type WBP to BS 6566:Part 8, Portland cement bonded particleboard to BS EN 633, non-asbestos mineral board), surface laminate to BS EN 438:Part 1, minimum VG (vertical grade), or BS 7332 grade G.	Regular cleaning with mild detergent.
20	High pressure laminate faced boards/panels to BS 4965. Core material to BS 4965 minimum durability class D3 (ie plywood to BS EN 636-3, particleboard to BS EN 312;Part 5, medium density fibreboard to BS EN 622-5, type MDF.H, tempered hardboard type HB.HLA 1 / 2 to BS EN 622-2), surface laminate to BS EN 438:Part 1, minimum VG (vertical grade), or BS 7332 grade G.	Regular cleaning with mild detergent.
15	High pressure laminate faced boards/panels to BS 4965. Core material to BS 4965 minimum durability class D2 (ie plywood to BS EN 636-1, particleboard to BS EN 312:Part 2/3, medium density fibreboard to BS EN 622-5, type MDF, standard hardboard type HB to BS EN 622-2, blockboard/laminboard of 5-ply construction, vermiculite based board), surface laminate to BS EN 438:Part 1, minimum VG (vertical grade), or BS 7332 grade G.	Regular cleaning with mild detergent.
U1	Unclassified, ie panels not to BS 4965, surface laminate not to BS EN 438:Part 1.	Unclassified.
Veneer/fabric/vinyl faced panels		
20	Factory faced boards/panels, class D4 core material as defined above (ie plywood type WBP to BS 6566:Part 8, Portland cement bonded particleboard to BS EN 633, non-asbestos mineral board).	Regular cleaning with mild detergent.
15	Factory faced boards/panels, class D3 core material as defined above (ie plywood to BS EN 636-3, particleboard to BS EN 312;Part 5, medium density fibreboard to BS EN 622-5, type MDF.H, tempered hardboard type HB.HLA 1 / 2 to BS EN 622-2).	Regular cleaning with mild detergent.
10	Factory faced boards/panels, class D2 core material as defined above (ie plywood to BS EN 636-1, particleboard to BS EN 312:Part 2/3, medium density fibreboard to BS EN 622-5, type MDF, standard hardboard type HB to BS EN 622-2, blockboard/laminboard of 5-ply construction, vermiculite based board).	Regular cleaning with mild detergent.
U1	Unclassified, ie panels less than class D2 above.	Unclassified.

For Adjustment factors, Assumptions, Notes, Key failure modes and Key durability issues please see overleaf.

2.11

Internal partitions –
timber based panels

LOCATIONS - General

SUB TYPES
Laminate faced panels
Veneer/fabric/vinyl faced panels

Notes

BS 5234:Part 2 defines four use categories for internal partitions (light, medium, heavy, severe duty), along with strength requirements and tests for each category. Tests include hard and soft body impact, stiffness and stability under loading.

Where a partition is required to support additional loads (such as lightweight fittings, wash basins, wall cupboards), BS 5234:Part 1 recommends that a sample of the partition is first tested using the methods in BS 5234:Part 2.

For gypsum based plasterboard, gypsum fibreboard, calcium silicate board, see HAPM Component Life Manual p. 2.30-2.31.

Key failure modes

Abrasion/peeling/delamination/fracture/tearing of facing material, moisture damage to core material (eg. curling/expansion/delamination/disintegration), impact, loss/pull out of fixings.

Key durability issues

Design for intended use (ie BS 5234 strength class), adequacy of support/fixings/stiffening, moisture resistance of core material/protection against moisture.

Adjustment factors

Partition system (including framework) Kitemarked to BS 5234:Part 1: +5 years (laminate faced boards/panels only).

Use in damp/humid environments, eg kitchens, bathrooms: -5 years (class D3 core material only).

Use in damp/humid environments, eg kitchens, bathrooms: -10 years (class D2 core material only).

Panels used as part of folding/sliding partition system: -5 years.

Assumptions

Grade HG laminates to BS 4965 should be used for heavy duty wall coverings, ie. locations where frequent impact is likely to occur.

Design and installation of partition system in accordance with BS 5234:Part 1.

Design for appropriate use category in accordance with BS 5234:Part 2. Partitions for use in office locations to be minimum 'medium duty'. Partitions for use in industrial or public circulation areas to be minimum heavy duty.

All laminate faced boards should be provided with a backing sheet on their reverse side. All unfaced edges sealed with lacquer, adhesive or a backing laminate.

All concealed metal channels/framing members to be minimum galvanized steel with 275g/m^2 zinc coating weight, or mill finished/anodized aluminium . Exposed metal framing members should have a life assessment at least equivalent to the boards/panels (see P.2.2 (framing) above).

Where partitions are built on platform floors and/or rely on suspended ceilings for their stability, the strength of these elements should be checked. The fixing of a partition to the underside of a suspended ceiling may require additional fixing supports provided above the suspended ceiling. On platform floors, partitions should be located at the positions recommended by the floor manufacturer and within the maximum recommended loadings. Otherwise, additional supports may be required.

Fixings to be at least galvanized steel or non ferrous and compatible with other metals used.

Surfaces likely to encounter moisture from splashes (eg around sinks) should be protected by tiling or other suitable material.

Boards/panels may be used as infill to separately provided framework, as the facing element of proprietary panels with a solid/framed core, or as linings to metal/timber studwork.

Suspended ceiling systems

Scope

This section provides data on the different types of suspended ceiling systems commonly used in non-domestic building types. It includes primary suspension grids, tile/tray infill panels, and interlocking metal plank/channel sections. In-situ plaster finishes are covered in Section 2 of the HAPM Component Life Manual and are excluded from this section. Open baffle and stretched fabric ceilings are also excluded

The following component sub-types are included within this section:

		Page
• Suspension grids:	Steel	2.12
	Aluminium	2.12
• Planks/channel sections:	Steel	2.13
	Aluminium	2.13
• Tiles/trays/infill panels	Steel	2.14
	Aluminium	2.14
	Gypsum fibreboard	2.14a
	Calcium silicate board	2.14a
	Gypsum based plasterboard	2.14a
	Mineral fibre	2.14b
	Rigid foam	2.14b
	Bonded vermiculite/perlite	2.14b

Standards cited

BS 970:

Part 1:1996 — Specification for wrought steels for mechanical and allied engineering purposes. General inspection and testing procedures and specific requirements for carbon, carbon manganese, alloy and stainless steels.

BS 1052:1980(1986) — Specification for mild steel wire for general engineering purposes.

BS 1230:

Part 1:1985(1994) — Gypsum plasterboard. Specification for plasterboard excluding materials submitted to secondary operations.

BS 1474: 1987 — Specification for wrought aluminium and aluminium alloys for general engineering purposes: bars, extruded round tubes and sections.

BS 1615:1987(1994) — Method for specifying anodic oxidation coatings on aluminium and it's coatings.

BS 4842: 1984(1991) — Specification for liquid organic coatings for application to aluminium alloy extrusions, sheet and preformed sections for external architectural purposes, and for the finish on aluminium alloy extrusions, sheet and preformed sections coated with liquid organic coatings.

BS 4921:1988 (1994) — Specification for sheradized coatings on iron or steel.

BS 6496:1984 (1991) — Specification for powder organic coatings for application and stoving to aluminium alloy extrusions, sheet and preformed sections for external architectural purposes, and for the finish on aluminium alloy extrusions, sheet and preformed sections coated with powder organic coatings.

BS 8290: — Suspended ceilings.

Part 1:1991 — Code of practice for design.

BS EN 485: — Aluminium and aluminium alloys. Sheet, strip and plate.

BS EN 10142:1991 — Specification for continuously hot dip zinc coated low carbon steel sheet and strip for cold forming: technical delivery conditions.

Other references/information sources

Department of Health: 'Health Technical Memorandum HTM 60: Ceilings'.

Suspended ceilings –
primary suspension grids

LOCATIONS - General

2 - Walling & Cladding Components

General	Description	Maintenance
	Steel	
35+	Austenitic stainless steel to BS 970:Part 1, grade 316 or 304.	None.
35+	Ferritic stainless steel to BS 970:Part 1, grade 430.	None.
35	Pre-galvanized steel to BS EN 10142, minimum 275g/m² zinc coating weight. Polyester powder coating to BS EN 10161 or liquid organic coating to BS 4842, minimum 40 micron coating thickness.	None.
30	Pre-galvanized steel to BS EN 10142, minimum 275g/m² zinc coating weight.	None.
25	Pre-galvanized steel to BS EN 10142, minimum 140g/m² zinc coating weight.	None.
U1	Unclassified, ie not to relevant BS/EN or less than above specification.	Unclassified.
	Aluminium	
35+	Aluminium alloy to BS 1474 (extrusions) or BS EN 485 (fabrications/sheet). Polyester powder coating to BS 6496 or liquid organic coating to BS 4842, minimum 40 micron coating thickness.	None.
35+	Aluminium alloy to BS 1474 (extrusions) or BS EN 485 (fabrications/sheet). Polyester powder coating to BS 6496 or liquid organic coating to BS 4842, minimum 25 micron coating thickness.	None.
35+	Aluminium alloy to BS 1474 (extrusions) or BS EN 485 (fabrications/sheet). Anodized to BS 1615, minimum 20 micron coating thickness.	None.
35	Aluminium alloy to BS 1474 (extrusions) or BS EN 485 (fabrications/sheet). Anodized to BS 1615, minimum 5 micron coating thickness.	None.
30	Aluminium alloy to BS 1474 (extrusions) or BS EN 485 (fabrications/sheet). Mill finished.	None.
U1	Unclassified, ie not to relevant BS/EN or less than above specification.	Unclassified.

For Adjustment factors, Assumptions, Key failure modes and Key durability issues please see overleaf.

2.12

Suspended ceilings –
primary suspension grids

LOCATIONS - General

Key failure modes

Impact/indentation/distortion, corrosion, chalking/ peeling/abrasion of decorative coatings, loss/pull out of fixings.

Key durability issues

Design for intended environment, adequacy of support/ fixings, type/thickness of metal coating.

Adjustment factors

Use in 'severe' environment (as defined in BS 8290:Part 1, and below): -10 years (except austenitic stainless steel).

Life limited to 10 years if plain steel fixings used.

Assumptions

Design and installation of suspended ceiling systems in accordance with BS 8290.

Design for appropriate environmental exposure category in accordance with BS 8290:Part 1. This standard defines two categories as follows:

Normal environment: conditions usually found in an occupied building where the temperature is within the range 10 degC to 30 degC and the relative humidity is regularly within the range 30% rh to 70% rh, and atmosphere does not contain acidic, alkaline or oxidizing solvent concentrations.

Severe environment: conditions that do not fall within the limits laid down for normal environments including ceilings installed in situations open to the external atmosphere.

For severe environments, coating thicknesses to be minimum 275g/m² for galvanized steel, and 20 microns for anodized aluminium. Additional protection may be required for very severe conditions such as swimming pool enclosures. Stainless steel suspension wires may be susceptible to stress corrosion in atmospheres (such as swimming pools) where chlorides occur.

Hangers and fixings (including any panel/tile clips) to be at least galvanized steel or non ferrous and compatible with other metals used. Galvanized steel wire hangers to be to BS 1052, minimum 2mm diameter. Galvanized steel wire is not recommended for 'severe' environments.

Fixing into structural ceiling/soffit to be adequate. Fixings into concrete to be expanding anchor type.

Suspended ceilings –
interlocking planks/channel sections

LOCATIONS - General

2 - Walling & Cladding Components

General	Description	Maintenance
	Steel	
30	Pre-galvanized steel to BS EN 10142, minimum 275g/m² zinc coating weight. Polyester powder coating to BS EN 10161 or liquid organic coating to BS 4842, minimum 40 micron coating thickness.	None.
25	Pre-galvanized steel to BS EN 10142, minimum 275g/m² zinc coating weight.	None.
20	Pre-galvanized steel to BS EN 10142, minimum 140g/m² zinc coating weight.	None.
U1	Unclassified, ie not to relevant BS/EN or less than above specification.	Unclassified.
	Aluminium	
35	Aluminium alloy to BS 1474 (extrusions) or BS EN 485 (fabrications/sheet). Polyester powder coating to BS 6496 or liquid organic coating to BS 4842, minimum 40 micron coating thickness.	None.
35	Aluminium alloy to BS 1474 (extrusions) or BS EN 485 (fabrications/sheet). Polyester powder coating to BS 6496 or liquid organic coating to BS 4842, minimum 25 micron coating thickness.	None.
35	Aluminium alloy to BS 1474 (extrusions) or BS EN 485 (fabrications/sheet). Anodized to BS 1615, minimum 20 micron coating thickness.	None.
30	Aluminium alloy to BS 1474 (extrusions) or BS EN 485 (fabrications/sheet). Anodized to BS 1615, minimum 5 micron coating thickness.	None.
25	Aluminium alloy to BS 1474 (extrusions) or BS EN 485 (fabrications/sheet). Mill finished.	None.
U1	Unclassified, ie not to relevant BS/EN or less than above specification.	Unclassified.

For Adjustment factors, Assumptions, Notes, Key failure modes and Key durability issues please see overleaf.

2.13

Suspended ceilings –
interlocking planks/channel sections

LOCATIONS - General

Adjustment factors

Use in 'severe' environment (as defined in BS 8290:Part 1, and below): -10 years.

Life limited to 10 years if plain steel fixings used.

Routine access to ceiling void (ie more than twice per year): -5 years.

Assumptions

Design and installation of suspended ceiling systems in accordance with BS 8290.

Planks/channel sections securely clipped to steel/ aluminium grid sections to relevant BS/EN as described on previous page.

Design for appropriate environmental exposure category in accordance with BS 8290:Part 1. This standard defines two categories as follows:

Normal environment: conditions usually found in an occupied building where the temperature is within the range 10 degC to 30 degC and the relative humidity is

regularly within the range 30% rh to 70% rh, and atmosphere does not contain acidic, alkaline or oxidizing solvent concentrations.

Severe environment: conditions that do not fall within the limits laid down for normal environments including ceilings installed in situations open to the external atmosphere.

For severe environments, coating thicknesses to be minimum 275g/m² for galvanized steel, and 20 microns for anodized aluminium. Additional protection may be required for very severe conditions such as swimming pool enclosures. Stainless steel suspension wires may be susceptible to stress corrosion in atmospheres (such as swimming pools) where chlorides occur.

Hangers and fixings (including any panel/tile clips) to be at least galvanized steel or non ferrous and compatible with other metals used. Galvanized steel wire hangers to be to BS 1052, minimum 2mm diameter. Galvanized steel wire is not recommended for 'severe' environments.

Fixing into structural ceiling/soffit to be adequate. Fixings into concrete to be expanding anchor type.

Notes

For gypsum based plasterboard, gypsum fibreboard, calcium silicate board, see HAPM Component Life Manual p. 2.30-2.31.

Key failure modes

Impact/indentation/distortion, corrosion, chalking/ peeling/ abrasion of decorative coatings, loss/pull out of fixings.

Key durability issues

Design for intended environment, adequacy of support/ fixings, type/thickness of metal coating.

Suspended ceilings –
tiles/trays/infill panels

LOCATIONS - General

BPG **2 - Walling & Cladding Components**

General	Description	Maintenance
Steel		
30	Pre-galvanized steel to BS EN 10142, minimum 275g/m² zinc coating weight. Polyester powder coating to BS EN 10161 or liquid organic coating to BS 4842, minimum 40 micron coating thickness.	None.
25	Pre-galvanized steel to BS EN 10142, minimum 275g/m² zinc coating weight.	None.
20	Pre-galvanized steel to BS EN 10142, minimum 140g/m² zinc coating weight.	None.
U1	Unclassified, ie not to relevant BS/EN or less than above specification.	Unclassified.
Aluminium		
35	Aluminium alloy to BS 1474 (extrusions) or BS EN 485 (fabrications/sheet). Polyester powder coating to BS 6496 or liquid organic coating to BS 4842, minimum 40 micron coating thickness.	None.
35	Aluminium alloy to BS 1474 (extrusions) or BS EN 485 (fabrications/sheet). Polyester powder coating to BS 6496 or liquid organic coating to BS 4842, minimum 25 micron coating thickness.	None.
35	Aluminium alloy to BS 1474 (extrusions) or BS EN 485 (fabrications/sheet). Anodized to BS 1615, minimum 20 micron coating thickness.	None.
30	Aluminium alloy to BS 1474 (extrusions) or BS EN 485 (fabrications/sheet). Anodized to BS 1615, minimum 5 micron coating thickness.	None.
25	Aluminium alloy to BS 1474 (extrusions) or BS EN 485 (fabrications/sheet). Mill finished.	None.
U1	Unclassified, ie not to relevant BS/EN or less than above specification.	Unclassified.

For Adjustment factors, Assumptions, Key failure modes and Key durability issues please see overleaf.

2.14

Suspended ceilings –
tiles/trays/infill panels

LOCATIONS - General

Adjustment factors

Use in 'severe' environment (as defined in BS 8290:Part 1, and below): -10 years.

Life limited to 10 years if plain steel fixings used.

Routine access to ceiling void (ie more than twice per year): -5 years.

Assumptions

Design and installation of suspended ceiling systems in accordance with BS 8290.

Tiles/trays/infill panels carried in steel/aluminium grid sections to relevant BS/EN as described on page 9.1.

Design for appropriate environmental exposure category in accordance with BS 8290:Part 1. This standard defines two categories as follows:

Normal environment: conditions usually found in an occupied building where the temperature is within the range 10 degC to 30 degC and the relative humidity is regularly within the range 30% rh to 70% rh, and atmosphere does not contain acidic, alkaline or oxidizing solvent concentrations.

Severe environment: conditions that do not fall within the limits laid down for normal environments including ceilings installed in situations open to the external atmosphere.

For severe environments, coating thicknesses to be minimum $275g/m^2$ for galvanized steel, and 20 microns for anodized aluminium. Additional protection may be required for very severe conditions such as swimming pool enclosures. Stainless steel suspension wires may be susceptible to stress corrosion in atmospheres (such as swimming pools) where chlorides occur.

Hangers and fixings (including any panel/tile clips) to be at least galvanized steel or non ferrous and compatible with other metals used. Galvanized steel wire hangers to be to BS 1052, minimum 2mm diameter. Galvanized steel wire is not recommended for 'severe' environments.

Fixing into structural ceiling/soffit to be adequate. Fixings into concrete to be expanding anchor type.

Key failure modes

Impact/indentation/distortion, corrosion, chalking/peeling/abrasion of decorative coatings, loss/pull out of fixings.

Key durability issues

Design for intended environment, adequacy of support/fixings, type/thickness of metal coating.

Suspended ceilings –
tiles/trays/infill panels

LOCATIONS - General

SUB TYPES:
Gypsum fibreboard
Calcium silicate board
Gypsum based plasterboard

BPG | 2 - Walling & Cladding Components

General	Description	Maintenance
	Gypsum fibreboard	
35	Gypsum fibreboard comprising a mixture of gypsum and cellulose fibre. BBA or other 3rd party approved for the particular application.	None.
30	Gypsum fibreboard comprising a mixture of gypsum and cellulose fibre. Approved by the manufacturer for the particular application.	None.
U1	Unclassified, ie board of unknown characteristics and/or not designated for proposed use.	Unclassified.
	Calcium silicate board	
35	Fibre-reinforced calcium silicate board. BBA or other 3rd party approved for the particular application.	None.
30	Fibre-reinforced calcium silicate board. Approved by the manufacturer for the particular application.	None.
U1	Unclassified, ie board of unknown characteristics and/or not designated for proposed use.	Unclassified.
	Gypsum based plasterboard	
30	Gypsum wallboard to BS 1230:Part 1, Type 3 (moisture resistant).	None.
25	Gypsum wallboard to BS 1230:Part 1, Types 1, 2, 4 or 5.	None.
U1	Unclassified, ie gypsum wallboard not to BS 1230:Part 1 or less than above specification.	Unclassified.

Adjustment factors

Routine access to ceiling void (ie more than twice per year): -10 years.

Assumptions

Materials to be suitable for the environment. Certain materials, including mineral fibre and some types of plasterboard, are unsuitable for use in damp or humid environments.

Reference should be made to manufacturers' guidance.

Cleaning in accordance with manufacturer's instructions.

Key failure modes

Impact/indentation/distortion, fracture/disintegration, tearing/ damage to surface finish.

Key durability issues

Frequency of disruption/access to ceiling void, environmental conditions.

2.14a

Suspended ceilings –
tiles/trays/infill panels

LOCATIONS - General

SUB TYPES
Mineral fibre
Rigid foam
Bonded
vermiculite/
perlite

2 - Walling & Cladding Components

General	Description	Maintenance
	Mineral fibre	
25	Resin bonded mineral wool, factory finished (eg with fabric, aluminum foil, glass tissue laminate). BBA or other 3rd party approved for the particular application.	None.
20	Resin bonded mineral wool, factory finished (eg with fabric, aluminum foil, glass tissue laminate).	None.
	Rigid foam	
20	Rigid urethane foam, factory finished (eg with fabric, aluminum foil, glass tissue laminate).	None.
15	Rigid urethane foam on melamine resin base. Self-finished.	None.
	Bonded vermiculite/perlite	
30	Bonded vermiculite.	None.
30	Bonded perlite.	None.

Key failure modes

Impact/indentation/distortion, fracture/disintegration, tearing/ damage to surface finish.

Key durability issues

Frequency of disruption/access to ceiling void, environmental conditions.

Adjustment factors

Routine access to ceiling void (ie more than twice per year): -10 years.

Assumptions

Materials to be suitable for the environment. Certain materials, including mineral fibre and some types of plasterboard, are unsuitable for use in damp or humid environments.

Reference should be made to manufacturers' guidance.

Cleaning in accordance with manufacturer's instructions.

2.14b

Roofing Components

Roofing Components

BPG

Rooflights

Scope

This section provides data on rooflights for flat and pitched roofs in non-domestic building types. It includes frames, preformed kerbs/upstands, and glazing materials including plastic profiled sheets for use with profiled metal roof decks. Sealed glazing units and domestic pitched roof windows are covered in the HAPM Component Life Manual and are excluded from this section.

The following component sub-types are included within this section:

		Page
Glazing:	GRP	3.1
	Polycarbonate	3.1
	Acrylic	3.1a
	PVC-u	3.1a
	Glass	3.1b
Glazing bars/frames:	Aluminium	3.2
	PVC-u	3.2
Preformed kerbs/upstands:	Aluminium	3.3
	Mild steel	3.3
	GRP	3.3

Standards cited

BS 729: 1971 (1994)	Specification for hot-dip galvanised coatings on iron and steel articles
BS 1449 Part 1: 1991	Steel plate, sheet and strip Carbon and carbon-manganese plate, sheet and strip
BS 1474: 1987	Specification for wrought aluminium and aluminium alloy for general engineering purposes : bars, extruded round tubes and sections
BS 1615: 1987 (1994)	Method for specifying anodic oxidation coatings on aluminium and its alloys
BS 2782 Part 10:	Methods of testing plastics Glass reinforced plastics
BS 3987: 1991 (1997)	Specification for anodic oxidation coatings on wrought aluminium for external architectural applications
BS 4154	Corrugated plastics translucent sheets made from thermo-setting polyester resin (glass fibre reinforced)
Part 1: 1985 (1994)	Specification for material and performance requirements
BS 4203	Extruded rigid PVC corrugated sheeting
Part 1: 1980 (1994)	Specification for performance requirements
BS 4842: 1984 (1991)	Specification for liquid organic coatings for application to aluminium alloy extrusions, sheet and preformed sections for external architectural purposes, and for the finish on aluminium alloy extrusions, sheet and preformed sections coated with liquid organic coatings.
BS 6206: 1981 (1994)	Specification for impact performance requirements for flat safety glass and safety plastics for use in buildings.
BS 6399	Loading for buildings
Part 1: 1996	Code of Practice for dead and imposed loads
Part 2: 1997	Code of Practice for wind loads
Part 3: 1988	Code of Practice for imposed roof loads
BS 6496: 1984 (1991)	Specification for powder organic coatings for application and stoving to aluminium alloy extrusions, sheet and preformed sections for external architectural purposes, and for the finish on aluminium alloy extrusions, sheet and preformed sections coated with powder organic coatings.
BS 7413: 1991	Specification for white PVC-u extruded hollow profiles with heat welded corner joints for plastic windows: materials type A.
BS 7722: 1994	Specification for surface covered PVC-u extruded hollow profiles with heat welded corner joints for plastics windows
BS EN 485: 1994/5	Aluminium and aluminium alloys. Sheet, strip and plate
BS EN 10142: 1991	Specification for continuously hot-dip zinc coated low carbon steel, sheet and strip for cold forming: technical delivery conditions.
BS EN 10147: 1992	Specification for continuously hot-dip zinc coated structural steel sheet and strip. Technical delivery conditions.

Other references/information sources

Association of Rooflight Manufacturers.

National Federation of Roofing Contractors (NFRC) Technical Bulletin 17: Guidance Note on the use of Rooflights with Profiled Cladding Systems.

HAPM Component Life Manual pages 4.11 - 4.12a, 4.13 (glazing sealants, sealed glazing units) and p.3.29 (pitched roof windows).

Rooflights - glazing

BPG 3 - Roofing Components

General	Description	Maintenance
GRP		
30	UV stabilised GRP sheet/section manufactured to BS 4154, minimum weight 2.44kg/m². Outer surface protected with factory applied fluoride-based film.	Periodic cleaning with warm water and mild (non-abrasive) detergent. Inspect fixings (and tighten if necessary) annually. Inspect gaskets annually, replace as necessary.
25	UV stabilised GRP sheet/section manufactured to BS 4154, minimum weight 2.44kg/m² for single skin applications, 1.83kg/m² (ie 1mm thick) for double/triple skin applications. Outer surface protected with factory applied UV protected film.	Periodic cleaning with warm water and mild (non-abrasive) detergent. Application of acrylic or polyester lacquer to restore sheet surface (typically between years 5 and 10, depending upon rate of surface erosion). Inspect fixings (and tighten if necessary) annually. Inspect gaskets annually, replace as necessary.
10	UV stabilised GRP sheet/section manufactured to BS 4154, minimum weight 1.83kg/m² (ie 1mm thick). Outer surface protected with polyester film/gel coat.	Periodic cleaning with warm water and mild (non-abrasive) detergent. Application of acrylic or polyester lacquer to restore sheet surface (typically between years 5 and 10, depending upon rate of surface erosion). Inspect fixings (and tighten if necessary) annually. Inspect gaskets annually, replace as necessary.
U1	Unclassified, ie GRP sheet/section not to BS 4154, and/or less than 1.83kg/m² weight/1mm thickness.	Unclassified.
Polycarbonate		
20	Polycarbonate sheet/section, minimum 1.5mm thick. UV protective surface film/coating to outer face.	Periodic cleaning with warm water and mild (non-abrasive) detergent. Application of acrylic or polyester lacquer to restore sheet surface (typically between years 5 and 10, depending upon rate of surface erosion). Inspect fixings (and tighten if necessary) annually. Inspect gaskets annually, replace as necessary.
15	Polycarbonate sheet/section, minimum 1.3mm thick. UV protective surface film/coating to outer face.	Periodic cleaning with warm water and mild (non-abrasive) detergent. Application of acrylic or polyester lacquer to restore sheet surface (typically between years 5 and 10, depending upon rate of surface erosion). Inspect fixings (and tighten if necessary) annually. Inspect gaskets annually, replace as necessary.
10	Polycarbonate sheet/section, less than 1.3mm thick. UV protective surface film/coating to outer face.	Periodic cleaning with warm water and mild (non-abrasive) detergent. Application of acrylic or polyester lacquer to restore sheet surface (typically between years 5 and 10, depending upon rate of surface erosion). Inspect fixings (and tighten if necessary) annually. Inspect gaskets annually, replace as necessary.
U1	Unclassified, ie polycarbonate sheet without UV protective surface film/coating to outer face.	Unclassified.

3.1 For Adjustment factors, Assumptions, Notes, Key failure modes and Key durability issues please see overleaf.

Rooflights - glazing

Key failure modes

UV degradation, weathering/surface erosion, impact, chafing/pull out around fixings, fracture/indentation due to imposed loading.

Key durability issues

UV protection, surface protection, material thickness, adequacy of fixings/supports, provision for thermal movement.

Notes

Some UV discolouration/yellowing can be expected to occur in most products within 5-10 years (including those with applied surface protection). However, recently introduced fluoride-based films are claimed to offer enhanced resistance to discolouration. GRP containing fire-retardant additives is likely to discolour rapidly.

Polycarbonate is prone to damage (embrittlement, stress cracking) by solvents and by the plasticizer in Plastisol coatings.

For applications where regular foot traffic is expected, a minimum material weight of 5.5kg/m³ is advisable, to ensure safety and resistance to damage (GRP only).

For sealed glazing units, see HAPM Component Life Manual p.4.13.

For glazing/frame sealants and gaskets, see HAPM Component Life Manual p.4.11-4.12a.

Domestic roof windows are included in the Spring 1998 update to the HAPM Component Life Manual.

Adjustment factors

Use in industrial/polluted/marine environment: -5 years.

Use in contact with Plastisol-coated steel, or rainwater wash from Plastisol coating: -5 years (Polycarbonate only).

Assumptions

The lives for double/triple skin products are based on the specification of the outer skin. The inner skin(s) may be of a lesser specification (eg not UV stabilised and/or surface protected).

Compatibility of sealant materials to be verified with glazing sheet manufacturer.

Glazing sheet thickness to be appropriate for the application.

Rooflights must not be painted over with an opaque covering.

Installation (including span, supports and fixing types/frequency) in accordance with manufacturer's instructions.

Thermoplastic (ie PVC-u, polycarbonate, acrylic) sheet requires oversized fixing holes to accommodate thermal movement.

All rooflight installations to be designed to accommodate site loading conditions in accordance with BS 6399: Part 1 (dead/imposed loads); Part 2 (wind loads); Part 3 (imposed roof loads).

Rooflights - glazing _____

BPG 3 - Roofing Components

General	Description	Maintenance
Acrylic		
20	Cast/extruded acrylic sheet/section, minimum 3mm thick.	Periodic cleaning with warm water and mild (non-abrasive) detergent. Application of acrylic or polyester lacquer to restore sheet surface (typically between years 5 and 10, depending upon rate of surface erosion). Inspect fixings (and tighten if necessary) annually. Inspect gaskets annually, replace as necessary.
15	Cast/extruded acrylic sheet/section, minimum 2mm thick.	Periodic cleaning with warm water and mild (non-abrasive) detergent. Application of acrylic or polyester lacquer to restore sheet surface (typically between years 5 and 10, depending upon rate of surface erosion). Inspect fixings (and tighten if necessary) annually. Inspect gaskets annually, replace as necessary.
U1	Unclassified, ie cast/extruded acrylic sheet, less than 2mm thick.	Unclassified.
PVC-u		
10	UV stabilised, extruded/thermoformed PVC-u sheet/section, minimum 1.5mm thick. Extruded sheets/sections to BS 4203:Part 1.	Periodic cleaning with warm water and mild (non-abrasive) detergent. Application of acrylic or polyester lacquer to restore sheet surface (typically between years 5 and 10, depending upon rate of surface erosion). Inspect fixings (and tighten if necessary) annually. Inspect gaskets annually, replace as necessary.
5	UV stabilised, extruded/thermoformed PVC-u sheet/section, minimum 1.3mm thick. Extruded sheets/sections to BS 4203:Part 1.	Periodic cleaning with warm water and mild (non-abrasive) detergent. Application of acrylic or polyester lacquer to restore sheet surface (typically between years 5 and 10, depending upon rate of surface erosion). Inspect fixings (and tighten if necessary) annually. Inspect gaskets annually, replace as necessary.
U1	Unclassified, ie PVC-u sheet/section, not UV stabilised, and/or less than 1.3mm thick, and/or extruded sheets not to BS 4203:Part 1.	Unclassified.

For Adjustment factors, Assumptions, Notes, Key failure modes and Key durability issues please see overleaf.

3.1a

Rooflights - glazing

Key failure modes

UV degradation, weathering/surface erosion, impact, chafing/pull out around fixings, fracture/indentation due to imposed loading.

Key durability issues

UV protection, surface protection, material thickness, adequacy of fixings/supports, provision for thermal movement.

Notes

Some UV discolouration/yellowing can be expected to occur in most products within 5-10 years (including those with applied surface protection). However, recently introduced fluoride-based films are claimed to offer enhanced resistance to discolouration.

For sealed glazing units, see HAPM Component Life Manual p.4.13.

For glazing/frame sealants and gaskets, see HAPM Component Life Manual p.4.11-4.12a.

Domestic roof windows are included in the Spring 1998 update to the HAPM Component Life Manual.

Adjustment factors

Use in industrial/polluted/marine environment: -5 years.

Factory-applied surface protection: +5 years (PVC-u only).

Assumptions

The lives for double/triple skin products are based on the specification of the outer skin. The inner skin(s) may be of a lesser specification (eg not UV stabilised and/or surface protected).

Compatibility of sealant materials to be verified with glazing sheet manufacturer.

Glazing sheet thickness to be appropriate for the application.

Rooflights must not be painted over with an opaque covering.

Installation (including span, supports and fixing types/frequency) in accordance with manufacturer's instructions.

Thermoplastic (ie PVC-u, polycarbonate, acrylic) sheet requires oversized fixing holes to accommodate thermal movement.

All rooflight installations to be designed to accommodate site loading conditions in accordance with BS 6399: Part 1 (dead/imposed loads); Part 2 (wind loads); Part 3 (imposed roof loads).

3.1a

Rooflights - glazing

LOCATIONS - General

3 - Roofing Components

General

Description

Glass

35+	Laminated glass to BS 6206.
35+	Toughened glass to BS 6206.
35	Wired glass to BS 6206.
U1	Unclassified, ie glass not to BS 6206 and/or .

Maintenance

Periodic cleaning with warm water and mild (non-abrasive) detergent. Inspect fixings (and tighten if necessary) annually. Inspect gaskets annually, replace as necessary.

Periodic cleaning with warm water and mild (non-abrasive) detergent. Inspect fixings (and tighten if necessary) annually. Inspect gaskets annually, replace as necessary.

Periodic cleaning with warm water and mild (non-abrasive) detergent. Inspect fixings (and tighten if necessary) annually. Inspect gaskets annually, replace as necessary.

Unclassified.

Key failure modes

Fracture due to imposed loading, cracking.

Key durability issues

Material thickness, adequacy of fixings/supports.

Notes

For sealed glazing units, see HAPM Component Life Manual p.4.13.

For glazing/frame sealants and gaskets, see HAPM Component Life Manual p.4.11-4.12a.

Domestic roof windows are included in the Spring 1998 update to the HAPM Component Life Manual.

Adjustment factors

None.

Assumptions

Compatibility of sealant materials to be verified with glazing sheet manufacturer.

Glazing sheet thickness to be appropriate for the application.

Installation (including span, supports and fixing types/frequency) in accordance with manufacturer's instructions.

All rooflight installations to be designed to accommodate site loading conditions in accordance with BS 6399: Part 1 (dead/imposed loads); Part 2 (wind loads); Part 3 (imposed roof loads).

3.1b

Rooflights -
glazing bars/frames

LOCATIONS - General

3 - Roofing Components

General	Description	Maintenance
Aluminium		
35	Aluminium to BS 1474 (extrusions)/BS EN 485 (fabrication/sheet). PVC/Plastisol coating, 200 microns nominal thickness.	Clean every 6 months with non-alkaline detergent (polluted/marine areas every 3 months). Lubricate ironmongery as required.
30	Aluminium to BS 1474 (extrusions)/BS EN 485 (fabrication/sheet). Acrylic, PVDF/PVF2), alkyd, polyester or silicone modified polyester coating, 25-50 microns nominal thickness.	Clean every 6 months with non-alkaline detergent (polluted/marine areas every 3 months). Coatings may require redecoration after 10-15 years. Lubricate ironmongery as required.
30	Aluminium to BS 1474 (extrusions)/BS EN 485 (fabrication/sheet). Anodised to BS 1615, minimum 25 micron coating.	Clean every 6 months with non-alkaline detergent (polluted/marine areas every 3 months). If painted, redecorate every 5 years. Lubricate ironmongery as required.
20	Aluminium to BS 1474 (extrusions)/BS EN 485 (fabrication/sheet). Anodised to BS 1615, minimum 10 micron coating.	Clean every 6 months with non-alkaline detergent (polluted/marine areas every 3 months). If painted, redecorate every 5 years. Lubricate ironmongery as required.
15	Aluminium to BS 1474 (extrusions)/BS EN 485 (fabrication/sheet). Mill finished.	Clean every 6 months with non-alkaline detergent (polluted/marine areas every 3 months). If painted, redecorate every 5 years. Lubricate ironmongery as required.
U1	Unclassified, ie aluminium, not to relevant BS.	Unclassified.
PVC-u		
20	White PVC-u hollow profile, multi chamber sections, internally reinforced as necessary with close fitting stainless steel, aluminium or pre-galvanized (G275) steel sections. Plastic to BS 7413 (modified plastics).	Clean every 6 months with non-alkaline detergent to maintain appearance. Lubricate ironmongery as required.
15	Film coloured PVC-u hollow profile, multi chamber sections, internally reinforced as necessary with close fitting stainless steel, aluminium or pre-galvanized (G275) steel sections. Plastic to BS 7413 (modified plastics). Colour film to comply with BS 7722.	Clean every 6 months with non-alkaline detergent to maintain appearance. Lubricate ironmongery as required.
U1	PVC-u, not to BS 7413. Any coloured PVC-u not complying with BS 7722.	Unclassified.

For Adjustment factors, Assumptions, Notes, Key failure modes and Key durability issues please see overleaf.

Rooflights -
glazing bars/frames

LOCATIONS - General

Adjustment factors

Industrial/polluted/marine environment: -5 years (aluminium only).

Life limited to 10 years if plain steel fixings used.

Assumptions

All fixings to be plated steel or non-ferrous, and compatible with frame materials.

Coatings to aluminium to BS 4842 (liquid organic), BS 6496 (powder), BS 3987 (anodising).

Installation in accordance with manufacturer's instructions.

Notes

Domestic roof windows are included in the Spring 1998 update to the HAPM Component Life Manual.

Solvent based cleaners should not be used. (PVC-u only).

Key failure modes

Corrosion, chalking/peeling of surface finish, misalignment/ wear/corrosion/fracture of ironmongery.

Key durability issues

Type/thickness of surface protection, maintenance of ironmongery.

Rooflights - preformed
kerbs & upstands

LOCATIONS - General

BPG 3 - Roofing Components

General	Description	Maintenance
Aluminium		
35	Aluminium to BS 1474 (extrusions)/BS EN 485 (fabrications/sheet). PVC or Plastisol coating, 200 microns nominal thickness.	Coatings may require redecoration after 10-15 years.
30	Aluminium to BS 1474 (extrusions)/BS EN 485 (fabrications/sheet). Acrylic, PVDF/PVF2, alkyd, polyester or silicone modified polyester coating, 25-50 microns nominal thickness.	Coatings may require redecoration after 10-15 years.
30	Aluminium to BS 1474 (extrusions)/BS EN 485 (fabrications/sheet). Anodized to BS 1615, minimum 25 micron coating.	If painted, redecorate every 5 years.
20	Aluminium to BS 1474 (extrusions)/BS EN 485 (fabrications/sheet). Anodized to BS 1615, minimum 10 micron coating.	If painted, redecorate every 5 years.
15	Aluminium to BS 1474 (extrusions)/BS EN 485 (fabrications/sheet). Mill finished.	If painted, redecorate every 5 years.
U1	Unclassified, ie aluminium, not to BS 1474 or BS EN 485.	Unclassified.
Mild steel		
30	Post-galvanized steel sheet to BS EN 10147, minimum 610g/m² zinc coating weight.	If painted, redecorate every 5 years.
30	Pre-galvanized steel sheet to BS 1449:Part 1/BS EN 10142, minimum 610g/m² zinc coating weight. Factory applied organic coating, 25-50 microns nominal thickness.	If painted, redecorate every 5 years.
25	Galvanized steel sheet to BS 1449:Part 1/BS EN 10142 or BS EN 10147. Minimum zinc coating 275g/m² (Z275). PVC or Plastisol coating, 200 microns nominal thickness.	Coatings may require redecoration after 10-15 years.
20	Galvanized steel sheet to BS 1449:Part 1/BS EN 10142 or BS EN 10147. Minimum zinc coating 275g/m² (Z275). Factory applied organic coating, 25-50 microns nominal thickness.	Coatings may require redecoration after 10-15 years.
10	Galvanized steel sheet to BS 1449:Part 1/BS EN 10142 or BS EN 10147. Minimum zinc coating 275g/m² (Z275).	If painted, redecorate every 5 years.
10	Mild steel, protected with an appropriate Micaceous Iron Oxide, Chlorinated Rubber or similar high performance finish to give a minimum dry film thickness of 250 microns.	Redecorate every 5 years.
U1	Unclassified, ie galvanized steel, zinc coating less than 275g/m² and/or not to BS 1449:Part 1/BS EN 10142/BS EN 10147.	Unclassified.
GRP		
20	GRP with a minimum corrugated laminate weight of 1.83kg/m² to BS 4154:Part 1. Protected with a polyester film coating to both sides and/or UV resistant resin rich layer of approximately 20 microns thick.	None.
15	GRP with a minimum corrugated laminate weight of 1.53kg/m² to BS 4154:Part 1. Protected with a polyester film coating to both sides and/or UV resistant resin rich layer of approximately 20 microns thick.	None.
U1	Unclassified, ie GRP with corrugated laminate weight less than 1.53kg/m² and/or no surface protection, and/or not fabricated in accordance with BS 4154: Part 1.	Unclassified.

3.3 For Adjustment factors, Assumptions, Key failure modes and Key durability issues please see overleaf.

Rooflights - preformed
kerbs & upstands

LOCATIONS - General

Adjustment factors

Industrial/polluted/marine environment: -5 years.

Life limited to 10 years if plain steel fixings used.

Assumptions

Fixings to be at least galvanized steel or non ferrous and compatible with other metals used.

Organic coatings include acrylic, PVDF/PVF2, alkyd, polyester or silicone modified polyester.

Hot dip galvanizing to BS 729.

All GRP laminate is opaque, with approximately 2% pigment and reinforced with chopped glass fibre mats or continuous rovings. All laminates are fully cured; minimum hardness to be 90% of hardness specified by resin supplier when tested by Barcol test as described in BS 2782:Part 10 method 1006.

Key failure modes

Corrosion, chalking/peeling of surface finish, UV degradation (GRP), weathering/surface erosion/ cracking/fracture (GRP).

Key durability issues

Type/thickness of surface protection, UV protection (GRP), material thickness (GRP).

3.3

Metal rainwater goods

Scope

This section provides data on metal rainwater goods for non-domestic building types. It includes eaves gutters, downpipes, proprietary siphonic systems and metal roof outlets. Plastic and other domestic rainwater goods are covered in the HAPM Component Life Manual and are excluded from this section.

The following component sub-types are included within this section:

		Page
• Eaves gutters & downpipes:	Stainless steel	3.4
	Mild steel	3.4
	Aluminium	3.4
	Copper	3.4a
	Zinc	3.4a
	Cast iron	3.4a
• Proprietary siphonic systems		3.5
• Roof outlets:	Cast gunmetal	3.6
	Nickel bronze	3.6
	Stainless steel	3.6
	Cast iron	3.6
	Aluminium	3.6a
	Mild steel	3.6a

Standards cited

BS 416	Discharge and ventilating pipes and fittings, sand-cast or spun in cast iron
Part 1: 1990	Specification for spigot and socket systems
BS 460: 1964 (1981)	Specification for cast iron rainwater goods
BS 729: 1971 (1994)	Specification for hot-dip galvanised coatings on iron and steel articles
BS 970	Specification for wrought steels for mechanical and allied engineering purposes
Part 1: 1996	General inspection and testing procedures and specific requirements for carbon, carbon manganese, alloy and stainless steels.
BS 1449	Steel plate, sheet and strip
Part 1:1991	Carbon and carbon-manganese plate, sheet and strip
Part 2:1983	Specification for stainless and heat resisting steel plate, sheet and strip
BS 1474: 1987	Specification for wrought aluminium and aluminium alloys for general engineering purposes: bars, extruded round tubes and sections
BS 1490: 1988	Specification for aluminium and aluminium alloy ingots and castings for general engineering purposes
BS 6496: 1984 (1991)	Specification for powder organic coatings for application and stoving to aluminium alloy extrusions, sheet and preformed sections for external architectural purposes, and for the finish on aluminium alloy extrusions, sheet and preformed sections coated with powder organic coatings
BS 8000	Workmanship on building sites
Part 13: 1989	Code of Practice for above ground drainage and sanitary appliances
BS EN 612: 1996	Eaves gutters and rainwater downpipes of metal sheet. Definitions, classifications and requirements
BS EN 100881: 1995	List of stainless steels
BS EN 10142: 1991	Specification for continuously hot-dip zinc coated low carbon steel, sheet and strip for cold forming: technical delivery conditions
BS EN 10147: 1992	Specification for continuously hot-dip zinc coated structural steel sheet and strip. Technical delivery conditions
BS EN 10214: 1995	Continuously hot-dip zinc-aluminium coated steel strip and sheet
BS EN 10215: 1995	Continuously hot-dip aluminium-zinc coated steel strip and sheet
ISO 6594	Cast iron drainage pipes and fittings - spigot series

Other references/information sources

Metal Gutter Manufacturers' Association: Metal Gutters Guide.

HAPM Component Life Manual pages 3.18 - 3.19 (domestic rainwater goods).

Metal rainwater goods -
eaves gutters & downpipes

LOCATIONS - General

Description

Stainless steel

General	Description	Maintenance
35+	Austenitic stainless steel sheet to BS EN 612 specification, BS 1449:Part 2/BS 970:Part 1 grade 316, or BS EN 10088-1 grade 1.4401.	Wash down every 6 months. Examine joints and seals annually, replace seals as necessary.
35+	Austenitic stainless steel sheet to BS EN 612 specification, BS 1449:Part 2/BS 970:Part 1 grade 304, or BS EN 10088-1 grade 1.4301.	Wash down every 6 months. Examine joints and seals annually, replace seals as necessary.
30	Ferritic stainless steel sheet to BS EN 612 specification, BS 1449:Part 2/BS 970:Part 1 grade 430, or BS EN 10088-1 grade 1.4510.	Wash down every 6 months. Examine joints and seals annually, replace seals as necessary.
U1	Unclassified, ie stainless steel, not to BS EN 612 specification.	Unclassified.

Mild steel

General	Description	Maintenance
30	Postgalvanized mild steel to BS EN 10147, minimum 610g/m² zinc coating weight.	Re-coating of bitumen paint to inside of gutter at 5 year intervals. Examine joints and seals annually, replace seals as necessary.
30	Pre-galvanized mild steel to BS 1449:Part 1/BS EN 10142, minimum 600g/m² zinc coating weight. Factory applied organic coating, 25-50 microns nominal thickness.	Re-coating of bitumen paint to inside of gutter at 5 year intervals. External coatings may require redecoration after 10-15 years. Examine joints and seals annually, replace seals as necessary.
25	Pre-galvanized mild steel to BS 1449:Part 1/BS EN 10142, minimum 275g/m² zinc coating weight. Factory applied PVC/Plastisol coating, 200 microns nominal thickness.	Re-coating of bitumen paint to inside of gutter at 5 year intervals. External coatings may require redecoration after 10-15 years. Examine joints and seals annually, replace seals as necessary.
20	Mild steel to BS EN 612 specification, hot dip zinc coated to BS EN 10142. Minimum coating mass of 275g/m². Organic coated on both sides, minimum thickness 25 microns (continuously coil coated), 60 microns (painted).	Paint at 5 year intervals (unless coil coated). Re-coating of bitumen paint to inside of gutter at 5 year intervals. Coil coatings may require redecoration after 10-15 years. Examine joints and seals annually, replace seals as necessary.
20	Mild steel to BS EN 612 specification, hot dip zinc-aluminium coated to BS EN 10214. Minimum coating mass of 225g/m². Organic coated on both sides, minimum thickness 25 microns (continuously coil coated), 60 microns (painted).	Paint at 5 year intervals (unless coil coated). Re-coating of bitumen paint to inside of gutter at 5 year intervals. Coil coatings may require redecoration after 10-15 years. Examine joints and seals annually, replace seals as necessary.
15	Mild steel to BS EN 612 specification, hot dip aluminium-zinc coated to BS EN 10215. Minimum coating mass of 150g/m². Organic coated on both sides, minimum thickness 25 microns (continuously coil coated), 60 microns (painted).	Paint at 5 year intervals (unless coil coated). Re-coating of bitumen paint to inside of gutter at 5 year intervals. Coil coatings may require redecoration after 10-15 years. Examine joints and seals annually, replace seals as necessary.
15	Mild steel to BS EN 612 specification, hot dip zinc coated to BS EN 10142. Minimum coating mass of 275g/m².	Re-coating of bitumen paint to inside of gutter at 5 year intervals. Examine joints and seals annually, replace seals as necessary.
U1	Unclassified, ie mild steel, not to BS EN 612 specification, and/or coating less than above specifications.	Unclassified.

For Adjustment factors, Assumptions, Key failure modes and Key durability issues please see overleaf.

3.4

Metal rainwater goods -
eaves gutters & downpipes

LOCATIONS - General

Key failure modes

Corrosion, chalking/peeling of surface finish, impact/indentation/distortion, misalignment/sagging, leaking joints.

Key durability issues

Type/thickness of surface protection, maintenance intervals (repainting, recoating of bitumen), material thickness, adequacy of fixings/supports, jointing method.

Adjustment factors

Use in industrial/marine environments: -10 years (mild steel only).

Use of 304 or 430 grade stainless steel in polluted/industrial/marine environments: -10 years.

Life limited to 10 years if plain steel fixings used.

BBA or other 3rd party certified systems: +5 years.

Assumptions

Minimum material thickness of metal sheet gutters and downpipes in accordance with BS EN 612.

No drainage from copper or lead roofs (or other parts) to come into contact with galvanized components.

All supports/fittings/brackets to be of the equivalent grade stainless steel. (Stainless steel only).

Installation in accordance with BS 8000:Part 13.

Spacing of supports/fixings in accordance with manufacturer's instructions.

Fixings at least galvanized steel or non ferrous and compatible with system (ie avoid use of stainless steel fixings with galvanized gutters/downpipes).

Galvanized steel gutters which are not protected with an organic coating should be coated on the inside with a bitumen paint or equivalent. After fixing, local areas of lost protection to be restored by application of a zinc-rich 'cold galvanizing' paint.

Downpipes located internally must have sealed joints.

Metal rainwater goods - eaves gutters & downpipes

LOCATIONS - General

BPG | **3 - Roofing Components**

General	Description	Maintenance
Aluminium		
35	Aluminium to BS EN 612 specification (sheet), BS 1490 (cast), BS 1474 (extruded). Polyester powder coated to BS 6496, minimum 50 micron coating thickness.	Redecorate at 15 years and every 5 years thereafter. Examine joints and seals annually, replace seals as necessary.
30	Aluminium to BS EN 612 specification (sheet), BS 1490 (cast), BS 1474 (extruded). Anodized to BS 1615, minimum 25 micron coating thickness.	Examine joints and seals annually, replace seals as necessary.
25	Aluminium to BS EN 612 specification (sheet), BS 1490 (cast), BS 1474 (extruded). Mill finished.	Examine joints and seals annually, replace seals as necessary.
U1	Unclassified, ie aluminium not to relevant BS.	Unclassified.
Copper		
35+	Copper sheet to BS EN 612 specification.	Examine joints and seals annually, replace seals as necessary.
U1	Unclassified, ie copper sheet, not to BS EN 612 specification.	Unclassified.
Zinc		
35	Zinc sheet to BS EN 612 specification.	Examine joints and seals annually, replace seals as necessary.
U1	Unclassified, ie zinc sheet, not to BS EN 612 specification.	Unclassified.
Cast iron		
35	Cast & ductile iron to BS 460/BS 416:Part 1.	Paint at 5 year intervals. Examine joints and seals annually, replace seals as necessary.
35	Cast & ductile iron to ISO 6594.	Paint at 5 year intervals. Examine joints and seals as necessary.
U1	Unclassified, ie cast & ductile iron not to BS 460/BS 416:Part 1/ISO 6594.	Unclassified.

For Adjustment factors, Assumptions, Notes, Key failure modes and Key durability issues please see overleaf.

3.4a

Metal rainwater goods -
eaves gutters & downpipes

LOCATIONS - General

BPG 3 - Roofing Components

Adjustment factors

Use in industrial/marine environments: -10 years (aluminium only).

Life limited to 10 years if plain steel fixings used.

Cast iron gutters epoxy coated internally: +5 years.

BBA or other 3rd party certified systems: +5 years.

Assumptions

Minimum material thickness of metal sheet gutters and downpipes in accordance with BS EN 612.

No drainage from copper or lead roofs (or other parts) to come into contact with aluminium components.

Installation in accordance with BS 8000:Part 13.

Spacing of supports/fixings in accordance with manufacturer's instructions.

Fixings at least galvanized steel or non ferrous and compatible with system (ie avoid use of stainless steel fixings with galvanized gutters/downpipes).

Downpipes located internally must have sealed joints.

Cast iron downpipes to be supported with brackets giving minimum 30mm clearance from the wall to facilitate painting.

Cast iron rainwater goods to be supplied factory primed with red oxide/zinc phosphate or other suitable primer.

Avoid run-off from aluminium roofing. (Cast iron only).

Notes

Aluminium is subject to significantly greater thermal movement than steel or cast iron; adequate provision must be made for movement when fixing/jointing.

Key failure modes

Corrosion, chalking/peeling of surface finish, impact/indentation/distortion, misalignment/sagging, leaking joints, fracture.

Key durability issues

Type/thickness of surface protection, maintenance intervals (repainting, recoating of bitumen), material thickness, adequacy of fixings/supports, jointing method.

3.4a

Metal rainwater goods -
eaves gutters & downpipes

LOCATIONS - General

3 - Roofing Components

General	Description		Maintenance
	Proprietary siphonic systems		
35+	Austenitic stainless steel to BS EN 612 specification, BS 1449:Part 2/BS 970:Part 1 grades 316 or 304, or BS EN 10088-1 grades 1.4401, 1.4301. Joints sealed with EPDM seals. BBA or other 3rd party certified system.		Examine and if necessary renew seals at 5-year intervals. Clear roof outlets of leaves/debris twice annually.
35	Austenitic stainless steel to BS EN 612 specification, BS 1449:Part 2/BS 970:Part 1 grades 316 or 304, or BS EN 10088-1 grades 1.4401, 1.4301. Joints sealed with EPDM seals.		Examine and if necessary renew seals at 5-year intervals. Clear roof outlets of leaves/debris twice annually.
35	Cast & ductile iron to BS 460/BS 416:Part 1, with EPDM seals.		Examine and if necessary renew seals at 5-year intervals. Clear roof outlets of leaves/debris twice annually. Paint at 5 year intervals if located externally.
35	Cast & ductile iron to ISO 6594, with EPDM seals.		Examine and if necessary renew seals at 5-year intervals. Clear roof outlets of leaves/debris twice annually. Paint at 5 year intervals if located externally.
30	Ferritic stainless steel to BS EN 612 specification, BS 1449:Part 2/BS 970:Part 1 grade 430, or BS EN 10088-1 grade 1.4510. Joints sealed with EPDM seals.		Examine and if necessary renew seals at 5-year intervals. Clear roof outlets of leaves/debris twice annually.
U1	Unclassified, ie stainless steel or cast iron, not to relevant BS.		Unclassified.

Adjustment factors

Life limited to 10 years if plain steel fixings used.

Assumptions

System must be designed in conjunction with manufacturer, using specialist computer software.

Cast iron supplied factory primed with red oxide/zinc phosphate or other suitable primer.

Avoid run-off from aluminium roofing. (Cast iron only).

Installation in accordance with BS 8000:Part 13.

Fixings at least galvanized steel or non ferrous and compatible with system.

Key failure modes

Corrosion, chalking/peeling of surface finish, impact/indentation/distortion, misalignment/sagging, loss of joint seal.

Key durability issues

Type/thickness of surface protection, maintenance intervals (eg repainting, clearing of outlets), material thickness, adequacy of fixings/supports, jointing method.

3.5

Metal rainwater goods -
roof outlets

LOCATIONS - General

3 - Roofing Components

General	Description	Maintenance
	Cast gunmetal	
35+	Outlet body/dome/grating of cast gunmetal to BS 1490.	Clear roof outlets of leaves/debris twice annually.
U1	Unclassified, ie cast gunmetal, not to BS 1490.	Unclassified.
	Nickel bronze	
35+	Outlet body/dome/grating of nickel bronze.	Clear roof outlets of leaves/debris twice annually.
	Stainless steel	
35+	Outlet body/dome/grating of 316 or 304 grade austenitic stainless steel to BS 970:Part 1.	Clear roof outlets of leaves/debris twice annually.
30	Outlet body/dome/grating of 430 grade ferritic stainless steel to BS 970:Part 1.	Clear roof outlets of leaves/debris twice annually.
U1	Unclassified, ie stainless steel, other than 316/304 grades, and/or not to relevant BS.	Unclassified.
	Cast iron	
35	Outlet body/dome/grating of cast iron to BS 460/BS 416:Part 1.	Clear roof outlets of leaves/debris twice annually. Paint at 5 year intervals.
35	Outlet body/dome/grating of cast iron-silicon alloy to BS 1490.	Clear roof outlets of leaves/debris twice annually. Paint at 5 year intervals.
U1	Unclassified, ie cast iron/cast iron-silicon alloy not to relevant BS.	Unclassified.

Adjustment factors

Use of 304 or 430 grade austenitic stainless steel in polluted/industrial/marine environments: -10 years.

Life limited to 10 years if plain steel fixings used.

Assumptions

All supports/fittings/brackets to be of the equivalent grade stainless steel. (Stainless steel only).

Installation in accordance with BS 8000:Part 13.

Fixings at least galvanized steel or non ferrous and compatible with system.

Supplied factory primed with red oxide/zinc phosphate or other suitable primer. (Cast iron only).

Key failure modes

Corrosion, chalking/peeling of surface finish, misalignment, leaking joints, fracture.

Key durability issues

Type/thickness of surface protection, maintenance intervals (repainting, clearing of outlets), adequacy of fixings/supports.

3.6

Metal rainwater goods –
roof outlets

LOCATIONS - General

BPG	3 - Roofing Components

Description

Aluminium

35+ Outlet body/dome/grating of cast aluminium to BS 1490. PVC/Plastisol coated, 200 microns nominal thickness.

35 Outlet body/dome/grating of cast aluminium to BS 1490. Acrylic, PVDF/PVF2, alkyd, polyester powder coated, 25-50 microns nominal thickness.

30 Outlet body/dome/grating of cast aluminium to BS 1490. Mill finished.

U1 Unclassified, ie cast aluminium not to BS 1490.

Mild steel

30 Outlet body/dome/grating of mild steel, hot dip galvanized after manufacture to BS 729, minimum 600g/m² zinc coating weight. PVC/Plastisol coated, 200 microns nominal thickness.

25 Outlet body/dome/grating of mild steel, hot dip galvanized after manufacture to BS 729, minimum 600g/m² zinc coating weight. Acrylic, PVDF/PVF2, alkyd, polyester powder coated, 25-50 microns nominal thickness.

20 Outlet body/dome/grating of mild steel, hot dip galvanized after manufacture to BS 729, minimum 600g/m² zinc coating weight

15 Outlet body/dome/grating of mild steel, hot dip galvanized after manufacture to BS 729, minimum 450g/m² zinc coating weight.

5 Outlet body/dome/grating of mild steel, hot dip galvanized after manufacture to BS 729, minimum 275g/m² zinc coating weight.

U1 Unclassified, ie hot dip galvanized steel, less than 275g/m² zinc coating weight.

Maintenance

Clear roof outlets of leaves/debris twice annually. Coatings may require redecoration after 10-15 years.

Clear roof outlets of leaves/debris twice annually. Coatings may require redecoration after 10-15 years.

Clear roof outlets of leaves/debris twice annually.

Unclassified.

Clear roof outlets of leaves/debris twice annually. Coatings may require redecoration after 10-15 years.

Clear roof outlets of leaves/debris twice annually. Coatings may require redecoration after 10-15 years.

Clear roof outlets of leaves/debris twice annually. Paint at 5 year intervals.

Clear roof outlets of leaves/debris twice annually. Paint at 5 year intervals.

Clear roof outlets of leaves/debris twice annually. Paint at 5 year intervals.

Unclassified.

Key failure modes

Corrosion, chalking/peeling of surface finish, misalignment, leaking joints.

Key durability issues

Type/thickness of surface protection, maintenance intervals (repainting, clearing of outlets), adequacy of fixings/supports.

Adjustment factors

Life limited to 10 years if plain steel fixings used.

Use in industrial/marine environments: -10 years.

Assumptions

No drainage from copper or lead roofs (or other parts) to come into contact with aluminium or galvanized components.

Installation in accordance with BS 8000:Part 13.

Fixings at least galvanized steel or non ferrous and compatible with system.

3.6a

Doors, Windows and Joinery Components

Doors, Windows and Joinery Components

External entrance doors

Scope

This section provides data on external entrance doors for use in non-domestic building types. It includes metal frame and infill panel doors, metal faced solid core doors, and frameless glass doors. Timber doors, PVC-u doors, metal doors for domestic use, and door frames are covered in Section 4 of the HAPM Component Life Manual and are excluded from this section. Specialised security doors are also excluded.

The following component sub-types are included within this section:

	Page
• Stainless steel	4.1
• Mild steel	4.1
• Aluminium	4.1a
• Frameless glass	4.1b

Standards cited

BS 729:1971(1994)	Specification for hot-dip galvanized coatings on iron and steel articles
BS 952:	Glass for glazing.
Part 1:1995	Classification.
BS 1449:	Steel plate, sheet and strip.
Part 2:1983	Specification for stainless and heat-resisting steel plate, sheet and strip.
BS 1474:1987	Specification for wrought aluminium and aluminium alloys for general engineering purposes: bars, extruded round tubes and sections.
BS 3987:1991(1997)	Specification for anodic oxidation coatings on wrought aluminium for external architectural applications.
BS 6262:	Code of practice for glazing for buildings.
Part 4:1982	Code of practice for safety. Human impact.
BS 6496:1984(1991)	Specification for powder organic coatings for application and stoving to hot-dip galvanized hot-rolled steel sections and preformed steel sheet for widows and associated external architectural purposes, and for the finish on galvanized steel sections and preformed sheet coated with powder organic coatings.
BS 6497:1984(1991)	Specification for powder organic coatings for application and stoving to hot-dip galvanized hot-rolled steel sections and preformed steel sheet for windows and external architectural purposes, and for the finish on galvanized steel sections and preformed sheet coated with powder organic coatings.
BS DD 171:1987	Guide to specifying performance requirements for hinged or pivoted doors (including test methods).
BS PD 6484:1979(1991)	Commentary on corrosion at bimetallic contacts and its alleviation
BS EN 572:1995	Glass in building. Basic soda lime silicate glass products.
BS EN 10088:2:1995	Technical delivery conditions for sheet/plate and strip for general purposes.
BS EN 10142:1991	Specification for continuously hot-dip zinc coated low carbon steel, sheet and strip for cold forming: technical delivery conditions.

Other references/information sources

Aluminium Window Association: 'Guidance in the handling, care, protection, fixing and maintenance of aluminium windows and doors'.

TYPE

External entrance doors_

LOCATIONS - General

SUB TYPES
Stainless steel
Mild steel

BPG | **4 - Doors, Windows & Joinery**

General	Description	Maintenance

Stainless steel

	Description	Maintenance
35+	Austenitic stainless steel framing to BS 1449:Part 2, grade 316 /BS EN 10088-2 grade 1.4401. Austenitic stainless steel or glazed infill panels.	Clean every 6 months with mild detergent. Redecorate every 5 years if site painted, after 20 years and every 5 years thereafter if powder coated. Replace glazing gaskets, weatherstripping etc as required. Tighten hinges as required.
35+	Door leaf faced both sides with minimum 2mm thick austenitic stainless steel to BS 1449:Part 2, grade 316 /BS EN 10088-2 grade 1.4401. Rigid mineral fibre or stainless steel grid core.	Clean every 6 months with mild detergent. Redecorate every 5 years if site painted, after 20 years and every 5 years thereafter if powder coated. Replace glazing gaskets, weatherstripping etc as required. Tighten hinges as required.
35	Door leaf faced both sides with minimum 2mm thick austenitic stainless steel to BS 1449:Part 2, grade 316 /BS EN 10088-2 grade 1.4401. Polyurethane foam core or honeycomb core of organic material.	Clean every 6 months with mild detergent. Redecorate every 5 years if site painted, after 20 years and every 5 years thereafter if powder coated. Replace glazing gaskets, weatherstripping etc as required. Tighten hinges as required.
U1	Unclassified, ie stainless steel not to BS 1449:Part 2 or BS EN 10088-2 and/or less than grade 316.	Unclassified.

Mild steel

	Description	Maintenance
35	Hot rolled mild steel frame, faced both sides with mild steel sheet of not less than 2.5mm thickness, hot dip galvanized after assembly to BS 729, minimum 1420g/m² zinc coating weight.	Clean every 6 months with mild detergent. Redecorate every 5 years if site painted, after 20 years and every 5 years thereafter if powder coated. Replace glazing gaskets, weatherstripping etc as required. Tighten hinges as required.
30	Hot rolled mild steel frame, hot dip galvanized to BS 729, minimum 1420g/m² zinc coating weight, faced both sides with minimum 2.5mm thick pre-galvanized mild steel sheet to BS 1449:Part 1/ BS EN 10142, minimum 600g/m² zinc coating weight. Polyester powder coated to BS 6497, minimum 60 micron nominal thickness.	Clean every 6 months with mild detergent. Redecorate after 20 years and every 5 years thereafter. Replace glazing gaskets, weatherstripping etc as required. Tighten hinges as required.
25	Cold formed pregalvanized mild steel frame to BS 1449:Part 1/ BS EN 10142, minimum 600g/m² zinc coating weight, with welded-on facing both sides of minimum 1.2mm thick pre-galvanized mild steel sheet to BS 1449:Part 1/ BS EN 10142, minimum 600g/m² zinc coating weight. Polyester powder coated to BS 6497, minimum 60 micron nominal thickness.	Clean every 6 months with mild detergent. Redecorate after 20 years and every 5 years thereafter. Replace glazing gaskets, weatherstripping etc as required. Tighten hinges as required.
20	Door leaf formed from mating pressed trays of pre-galvanized mild steel sheet to BS 1449:Part 1/ BS EN 10142. Honeycomb, rigid mineral wool or rigid foam core. Polyester powder coated to BS 6497, minimum 60 micron nominal thickness.	Clean every 6 months with mild detergent. Redecorate after 20 years and every 5 years thereafter. Replace glazing gaskets, weatherstripping etc as required. Tighten hinges as required.
15	Door leaf formed from mating pressed trays of pre-galvanized mild steel sheet to BS 1449:Part 1/ BS EN 10142. Hollow core. Polyester powder coated to BS 6497, minimum 60 micron nominal thickness.	Clean every 6 months with mild detergent. Redecorate after 20 years and every 5 years thereafter. Replace glazing gaskets, weatherstripping etc as required. Tighten hinges as required.
U1	Unclassified, ie galvanized mild steel not to BS 1449, BS EN 10142 or BS 729, and/or galvanizing less than that specified above.	Unclassified

4.1 | **For Adjustment factors, Assumptions, Notes, Key failure modes and Key durability issues please see overleaf.**

External entrance doors_

BPG 4 - Doors, Windows & Joinery

Adjustment factors

Industrial/polluted/marine environment: -5 years (except 316 grade austenitic stainless steel).

Door leaf which is part of a revolving or automated sliding door: +5 years.

Doors supplied primed for site application of paint finish: -5 years.

Assumptions

Installation (including attachment of ironmongery, frequency, type and fixing of hinges) in strict accordance with manufacturer's instructions.

Where used within the door leaf, steel/stainless steel framing is of welded construction.

Infill panels to framed doors assumed to be either glazed or of the same material/coating specification as the frame.

Glazing to BS 6262.

Hinges are either welded or, if bolted, door is supplied with reinforcing strips at hinge positions.

Hinged doors closing against rebates to comply with the relevant performance tests of BS DD 171.

All fixings to be plated steel or non-ferrous. Stainless steel ironmongery to be fitted with stainless steel fixings.

Where two or more metals are used in the door construction, they should be compatible, ie to prevent galvanic corrosion. For further detailed guidance, see BS PD 6484.

With galvanized hollow sections it is essential to provide holes for venting and drainage, and to ensure that the internal surfaces are fully coated (steel only).

Vulnerable facings should be provided with kick-plates or similar protection should be provided in areas of heavy use.

Notes

The above descriptions are indicative only; a wide variety of construction types and surface finishes are available, particularly for high security or fire-protection applications.

For door frames, see HAPM Component Life Manual p.4.2-4.2d.

For domestic entrance doors (including timber and PVC-u), see HAPM Component Life Manual p.4.1-4.1h.

Key failure modes

Mechanical damage (impact, abrasion), corrosion, chalking/peeling of surface finish.

Key durability issues

Base metal, type and thickness of surface protection, exposure conditions, strength of fixings between frame and door leaf.

4.1

External entrance doors

4 - Doors, Windows & Joinery

General	Description		Maintenance
	Aluminium		
35	Extruded aluminium alloy framing to BS 1474, with austenitic stainless steel drawn skin to BS 1449:Part 2, grade 316 /BS EN 10088-2 grade 1.4401. Compatible hardware and fixings.		Clean every 6 months with non-alkaline detergent (3 months in marine/polluted environment). If painted, redecorate every 5 years. Adjust and lubricate ironmongery as required. Replace draughtstrips and gaskets every 10 years. Tighten hinges as required.
25	Extruded aluminium alloy framing to BS 1474, anodized to BS 3987, minimum 25 micron coating. Polyester powder coated to BS 6496, minimum 40 micron thickness. Compatible hardware and fixings.		Clean every 6 months with non-alkaline detergent (3 months in marine/polluted environment). Adjust and lubricate ironmongery as required. Replace draughtstrips and gaskets every 10 years. Tighten hinges as required.
20	Extruded aluminium alloy framing to BS 1474, anodized to BS 3987, minimum 25 micron coating. Compatible hardware and fixings.		Clean every 6 months with non-alkaline detergent (3 months in marine/polluted environment). If painted, redecorate every 5 years. Adjust and lubricate ironmongery as required. Replace draughtstrips and gaskets every 10 years. Tighten hinges as required.
20	Extruded aluminium alloy framing to BS 1474, polyester powder coated to BS 6496, minimum 40 micron thickness. Compatible hardware and fixings.		Clean every 6 months with non-alkaline detergent (3 months in marine/polluted environment). Adjust and lubricate ironmongery as required. Replace draughtstrips and gaskets every 10 years. Tighten hinges as required.
10	Extruded aluminium alloy framing to BS 1474. Mill finished. Compatible hardware and fixings.		Clean every 6 months with non-alkaline detergent (3 months in marine/polluted environment). If painted, redecorate every 5 years. Adjust and lubricate ironmongery as required. Replace draughtstrips and gaskets every 10 years. Tighten hinges as required.
U1	Unclassified, ie extruded aluminium alloy not to BS 1474.		Unclassified.

For Adjustment factors, Assumptions, Notes, Key failure modes and Key durability issues please see overleaf.

Adjustment factors

Industrial/polluted/marine environment: -5 years.

Organic coating protected during delivery and installation by self-adhesive tape: -5 years.

Door leaf which is part of a revolving or automated sliding door: +5 years.

Assumptions

Installation (including attachment of ironmongery, frequency, type and fixing of hinges) in strict accordance with manufacturer's instructions.

Any frame fixing screws/bolts/nuts to be of austenitic stainless steel.

All fixings to be plated steel or non-ferrous. Stainless steel ironmongery to be fitted with stainless steel fixings.

Infill panels to framed doors assumed to be either glazed or of the same material/coating specification as the frame.

Coated aluminium is likely to require recoating after 12 years (liquid organic coating) or 15 years (polyester powder coating).

Hinged doors closing against rebates to comply with the relevant performance tests of BS DD 171.

In external/damp locations, avoid direct contact between aluminium alloys and timber treated with copper, zinc or mercury based preservatives, Oak, Sweet Chestnut, Douglas Fir, Western Red Cedar, copper alloys (or rainwater run off from), concrete, mortar or soil.

Where two or more metals are used in the door construction, they should be compatible, ie to prevent galvanic corrosion. For further detailed guidance, see BS PD 6484.

Vulnerable facings should be provided with kick-plates or similar protection should be provided in areas of heavy use.

Notes

Self-adhesive tape may pull part of the coating off when it is removed following completion of installation.

For door frames, see HAPM Component Life Manual p.4.2-4.2d.

For domestic entrance doors (including timber and PVC-u), see HAPM Component Life Manual p.4.1-4.1h.

Key failure modes

Mechanical damage (impact, abrasion), chalking/peeling of surface finish, corrosion.

Key durability issues

Type and thickness of surface protection, exposure conditions, strength of fixings between frame and door leaf.

4.1a

BPG | **4 - Doors, Windows & Joinery**

Description

Maintenance

Frameless glass

General		Maintenance
35+	Door leaf formed from a single sheet of heat soaked thermally toughened glass to BS 952: Part 1, BS EN 527 and BS 6262, of thickness appropriate to size and loading, with top and bottom members of grade 316 austenitic stainless steel to BS 1449:Part 2, grade 316 /BS EN 10088-2 grade 1.4401, or of bronze.	Lubrication and maintenance of door closer and hinge mechanisms as recommended by manufacturer.
30	Door leaf formed from a single sheet of heat soaked thermally toughened glass to BS 952: Part 1, BS EN 527 and BS 6262, of thickness appropriate to size and loading, with top and bottom members of aluminium alloy extrusions integrally clad with grade 316 austenitic stainless steel to BS 1449:Part 2, grade 316 /BS EN 10088-2 grade 1.4401.	Lubrication and maintenance of door closer and hinge mechanisms as recommended by manufacturer.
25	Door leaf formed from a single sheet of heat soaked thermally toughened glass to BS 952: Part 1, BS EN 527 and BS 6262, of thickness appropriate to size and loading, with top and bottom members of brass.	Lubrication and maintenance of door closer and hinge mechanisms as recommended by manufacturer.
20	Door leaf formed from a single sheet of heat soaked thermally toughened glass to BS 952: Part 1, BS EN 527 and BS 6262, of thickness appropriate to size and loading, with top and bottom members of aluminium alloy extrusions, polyester powder coated to BS 6496, minimum 40 micron thickness.	Lubrication and maintenance of door closer and hinge mechanisms as recommended by manufacturer.
U1	Unclassified, ie glass not to BS 952:Part 1, BS EN 572 and/or BS 6262.	Unclassified.

Adjustment factors

Glass not heat soaked: -10 years.

Assumptions

BS 6262 specifies the impact performance of three classes of safety glass. BS 6262:Part 4 requires that where the glass door leaf is more than 900mm wide, the glass should conform to at least class B. For door leaves less than 900mm wide, glass conforming to at least class C must be specified.

Door leaves are used in conjunction with floor-mounted door closers.

Installation (including holes, notches and mountings, attachment of ironmongery) in strict accordance with manufacturer's instructions.

Notes

The presence of nickel sulphide inclusions in toughened glass can lead to the risk of 'spontaneous breakage' during normal use. It is widely recognised that heat soaking the glass after toughening, which converts the nickel sulphide to a more stable form, significantly reduces this risk.

Key failure modes

Fracture, cracking, scratching, corrosion of top/bottom members, failure of hinges.

Key durability issues

Glass specification, intensity of use.

4.1b

Industrial doors (external)

Scope

This section provides data on large industrial doors for external use in non-domestic building types. It includes doors made up from horizontal or vertical metal sections, metal frames and infill panels, and also fabric-based traffic doors. Flexible plastic curtains are excluded from this section, as are security grilles and shutters. Door frames, guide rails and operating gear are not considered in detail.

The following component sub-types are included within this section:

	Page
• Rolling shutter	4.2
• Folding shutter (horizontal)	4.2a
• Folding shutter (horizontal)	4.2b
• Section overhead - frameless sections	4.2c
• Sectional overhead - frames	4.2d
• Sectional overhead - metal infill panels	4.2e
• Folding/sliding sectional	4.2f
• Fast action traffic doors	

Standards cited

BS 970	Specification for wrought steels and mechanical and allied engineering purposes
Part 1: 1996	General inspection and testing procedures and specific requirements for carbon, carbon manganese, alloy and stainless steels
BS 1449	Steel plate, sheet and strip
Part 1: 1991	Carbon and carbon-manganese plate, sheet and strip
Part 2: 1983	Specification for stainless and heat resisting steel plate, sheet and strip
BS 1474: 1987	Specification for wrought aluminium and aluminium alloy for general engineering purposes : bars, extruded round tubes and sections
BS 1615: 1987 (1994)	Method for specifying anodic oxidation coatings on aluminium and its alloys
BS 7079: (various dates)	Preparation of steel substrates before application of paints and related products
BS EN 485: 1994	Aluminium and aluminium alloys. Sheet, strip and plate
BS EN 10142: 1991	Specification for continuously hot-dip zinc coated low carbon steel sheet and strip for cold forming: technical delivery conditions
BS EN 10147: 1992	Specification for continuously hot-dip zinc coated structural steel sheet and strip. Technical delivery conditions

Industrial doors (external)

BPG 4 - Doors, Windows & Joinery

General — Rolling shutter - steel

	Description	Maintenance
30	Austenitic stainless steel lath shutter curtain to BS 1449:Part 2 or to BS 970:Part 1, Grades 316 or 304.	Clean every 6 months with non-alkaline detergent (polluted/ marine areas every 3 months). Lubrication of joints/bearings in accordance with manufacturer's recommendations, or cyclical maintenance carried out under contract with manufacturer.
20	Post galvanized steel lath shutter curtain to BS EN 10147, minimum 600g/m² zinc coating weight.	Clean every 6 months with non-alkaline detergent (polluted/ marine areas every 3 months). If painted, redecorate every 5 years. Lubrication of joints/bearings in accordance with manufacturer's recommendations, or cyclical maintenance carried out under contract with manufacturer.
15	Post galvanized steel lath shutter curtain to BS EN 10147, minimum 450g/m² zinc coating weight.	Clean every 6 months with non-alkaline detergent (polluted/ marine areas every 3 months). If painted, redecorate every 5 years. Lubrication of joints/bearings in accordance with manufacturer's recommendations, or cyclical maintenance carried out under contract with manufacturer.
15	Pre-galvanized steel lath shutter curtain to BS 1449:Part 1 or BS EN 10142 minimum 600g/m² zinc coating weight.	Clean every 6 months with non-alkaline detergent (polluted/ marine areas every 3 months). If painted, redecorate every 5 years. Lubrication of joints/bearings in accordance with manufacturer's recommendations, or cyclical maintenance carried out under contract with manufacturer.
15	Galvanized steel lath shutter curtain to BS 1449:Part 1/BS EN 10142 or BS EN 10147, minimum 275g/m2. PVC/Plastisol coating, 200 microns nominal thickness.	Clean every 6 months with non-alkaline detergent (polluted/ marine areas every 3 months). Coatings may require redecoration after 10-15 years. Lubrication of joints/bearings in accordance with manufacturer's recommendations, or cyclical maintenance carried out under contract with manufacturer.
10	Galvanized steel lath shutter curtain to BS 1449:Part 1/BS EN 10142 or BS EN 10147, minimum zinc coating 275g/m2. Acrylic, PVDF/PVF2, alkyd, polyester or silicone modified polyester coating, 25-50 microns nominal thickness.	Clean every 6 months with non-alkaline detergent (polluted/ marine areas every 3 months). Coatings may require redecoration after 10-15 years. Lubrication of joints/bearings in accordance with manufacturer's recommendations, or cyclical maintenance carried out under contract with manufacturer.
5	Galvanized steel lath shutter curtain to BS 1449:Part 1/BS EN 10142 or BS EN 10147, minimum zinc coating 275g/m².	Clean every 6 months with non-alkaline detergent (polluted/ marine areas every 3 months). If painted, redecorate every 5 years. Lubrication of joints/bearings in accordance with manufacturer's recommendations, or cyclical maintenance carried out under contract with manufacturer.
5	Mild steel lath shutter curtain, blast cleaned after any cutting, welding or drilling to BS 7079 'second quality' (equivalent to SA 2.5) and protected with an appropriate Micaceous Iron Oxide, Chlorinated Rubber or similar high performance finish to give a minimum dry film thickness of 250 microns.	Clean every 6 months with non-alkaline detergent (polluted/ marine areas every 3 months). Redecorate every 5 years. Lubrication of joints/bearings in accordance with manufacturer's recommendations, or cyclical maintenance carried out under contract with manufacturer.
U1	Unclassified, ie mild steel 'factory primed', steel not to relevant BS.	Unclassified.

4.2 — For Adjustment factors, Assumptions, Notes, Key failure modes and Key durability issues please see overleaf.

Industrial doors (external)

LOCATIONS - General

Adjustment factors

Industrial/polluted/marine environment: -5 years (except 316 grade austenitic stainless steel).

Guide rails, channels, supports, etc. 'factory primed' or less than 275g/m² galvanized steel: life limited to 10 years.

Assumptions

Installation in accordance with manufacturer's instructions.

Curtain thickness to be appropriate for required opening width. Guidance on selection should be obtained from manufacturer.

Guide rails, channels, supports, etc. to be minimum 275g/m² galvanized steel.

Fixings to be at least galvanized steel or non ferrous and compatible with other metals used.

Repair/replacement of opening gear and ironmongery as required.

Door operation may be manual, geared hoist or electrical motor driven.

Galvanized steel/aluminium hoods or fascias - lives as for shutter curtains.

Notes

Some fading and chalking of surface coatings may occur within the lives quoted.

A number of manufacturers offer maintenance contracts for the cyclical maintenance and repair of industrial doors.

Key failure modes

Corrosion, scratching/chalking/peeling of surface finish, impact/indentation/ distortion, misalignment/binding/ defective operation, failure of operating gear.

Key durability issues

Type/thickness of surface protection, curtain thickness/ profile, fixing/support/alignment of guide rails, maintenance frequency, intensity of use, protection against impact.

BPG

4 - Doors, Windows & Joinery

General

Description

Rolling shutter - aluminium

Maintenance

25 | Aluminium lath shutter curtain, to BS 1474/BS EN 485. PVC/Plastisol coating, 200 microns nominal thickness. | Clean every 6 months with non-alkaline detergent (polluted/ marine areas every 3 months). Coatings may require redecoration after 10-15 years. Lubrication of joints/bearings in accordance with manufacturer's recommendations, or cyclical maintenance carried out under contract with manufacturer.

20 | Aluminium lath shutter curtain to BS 1474/BS EN 485. Acrylic, PVDF/ PVF2, alkyd, polyester or silicone modified polyester coating, 25-50 microns nominal thickness. | Clean every 6 months with non-alkaline detergent (polluted/ marine areas every 3 months). Coatings may require redecoration after 10-15 years. Lubrication of joints/bearings in accordance with manufacturer's recommendations, or cyclical maintenance carried out under contract with manufacturer.

20 | Anodised aluminium lath shutter curtain to BS 1474/BS EN 485. Minimum 25 micron coating. | Clean every 6 months with non-alkaline detergent (polluted/ marine areas every 3 months). If painted, redecorate every 5 years. Lubrication of joints/bearings in accordance with manufacturer's recommendations, or cyclical maintenance carried out under contract with manufacturer.

10 | Anodised aluminium lath shutter curtain to BS 1474/BS EN 485. Minimum 10 micron coating. | Clean every 6 months with non-alkaline detergent (polluted/ marine areas every 3 months). If painted, redecorate every 5 years. Lubrication of joints/bearings in accordance with manufacturer's recommendations, or cyclical maintenance carried out under contract with manufacturer.

5 | Mill finished aluminium lath shutter curtain to BS 1474/BS EN 485. | Clean every 6 months with non-alkaline detergent (polluted/ marine areas every 3 months). If painted, redecorate every 5 years. Lubrication of joints/bearings in accordance with manufacturer's recommendations, or cyclical maintenance carried out under contract with manufacturer.

U1 | Unclassified, ie aluminium not to relevant BS. | Unclassified.

For Adjustment factors, Assumptions, Notes, Key failure modes and Key durability issues please see overleaf.

4.2a

Industrial doors (external)

LOCATIONS - General

BPG 4 - Doors, Windows & Joinery

Adjustment factors

Industrial/polluted/marine environment: -5 years.

Guide rails, channels, supports, etc. 'factory primed' or less than 275g/m² galvanized steel: life limited to 10 years.

Assumptions

Installation in accordance with manufacturer's instructions.

Curtain thickness to be appropriate for required opening width. Guidance on selection should be obtained from manufacturer.

Guide rails, channels, supports, etc. to be minimum 275g/m² galvanized steel.

Fixings to be at least galvanized steel or non ferrous and compatible with other metals used.

Repair/replacement of opening gear and ironmongery as required.

Door operation may be manual, geared hoist or electrical motor driven.

Galvanized steel/aluminium hoods or fascias - lives as for shutter curtains.

Notes

Some fading and chalking of surface coatings may occur within the lives quoted.

A number of manufacturers offer maintenance contracts for the cyclical maintenance and repair of industrial doors.

Key failure modes

Corrosion, scratching/chalking/peeling of surface finish, impact/indentation/ distortion, misalignment/binding/ defective operation, failure of operating gear.

Key durability issues

Type/thickness of surface protection, curtain thickness/ profile, fixing/support/alignment of guide rails, maintenance frequency, intensity of use, protection against impact.

BPG | **4 - Doors, Windows & Joinery**

General	Description	Maintenance
	Folding shutter (horizontal) - steel	
20	Post galvanized steel shutter leaves to BS EN 10147, minimum 600g/m² zinc coating weight.	Clean every 6 months with non-alkaline detergent (polluted/marine areas every 3 months). If painted, redecorate every 5 years. Lubrication of joints/bearings in accordance with manufacturer's recommendations, or cyclical maintenance carried out under contract with manufacturer.
15	Post galvanized steel shutter leaves to BS EN 10147, minimum 450g/m² zinc coating weight.	Clean every 6 months with non-alkaline detergent (polluted/marine areas every 3 months). If painted, redecorate every 5 years. Lubrication of joints/bearings in accordance with manufacturer's recommendations, or cyclical maintenance carried out under contract with manufacturer.
15	Pre-galvanized steel shutter leaves to BS 1449:Part 1 or BS EN 10142 minimum 600g/m² zinc coating weight.	Clean every 6 months with non-alkaline detergent (polluted/marine areas every 3 months). If painted, redecorate every 5 years. Lubrication of joints/bearings in accordance with manufacturer's recommendations, or cyclical maintenance carried out under contract with manufacturer.
15	Galvanized steel shutter leaves to BS 1449:Part 1/BS EN 10142 or BS EN 10147, minimum zinc coating 275g/m². PVC/Plastisol coating, 200 microns nominal thickness.	Clean every 6 months with non-alkaline detergent (polluted/marine areas every 3 months). Coatings may require redecoration after 10-15 years. Lubrication of joints/bearings in accordance with manufacturer's recommendations, or cyclical maintenance carried out under contract with manufacturer.
10	Galvanized steel shutter leaves to BS 1449:Part 1/BS EN 10142 or BS EN 10147, minimum zinc coating 275g/m². Acrylic, PVDF/PVF2, alkyd, polyester or silicone modified polyester coating, 25-50 microns nominal thickness.	Clean every 6 months with non-alkaline detergent (polluted/marine areas every 3 months). Coatings may require redecoration after 10-15 years. Lubrication of joints/bearings in accordance with manufacturer's recommendations, or cyclical maintenance carried out under contract with manufacturer.
5	Galvanized steel shutter leaves to BS 1449:Part 1/BS EN 10142 or BS EN 10147, minimum zinc coating 275g/m².	Clean every 6 months with non-alkaline detergent (polluted/marine areas every 3 months). If painted, redecorate every 5 years. Lubrication of joints/bearings in accordance with manufacturer's recommendations, or cyclical maintenance carried out under contract with manufacturer.
5	Mild steel shutter leaves, blast cleaned after any cutting, welding or drilling to BS 7079 'second quality' (equivalent to SA 2.5) and protected with an appropriate Micaceous Iron Oxide, Chlorinated Rubber or similar high performance finish to give a minimum dry film thickness of 250 microns.	Clean every 6 months with non-alkaline detergent (polluted/marine areas every 3 months). Redecorate every 5 years. Lubrication of joints/bearings in accordance with manufacturer's recommendations, or cyclical maintenance carried out under contract with manufacturer.
U1	Unclassified, ie mild steel 'factory primed', steel not to relevant BS.	Unclassified.

4.2b For Adjustment factors, Assumptions, Notes, Key failure modes and Key durability issues please see overleaf.

Industrial doors (external)

LOCATIONS - General

BPG   4 - Doors, Windows & Joinery

Adjustment factors

Industrial/polluted/marine environment: -5 years.

Guide rails, channels, supports, lattice gates etc. 'factory primed' or less than 275g/m² galvanized steel: life limited to 10 years.

Assumptions

Installation in accordance with manufacturer's instructions.

Leaf thickness to be appropriate for required opening size. Guidance on selection should be obtained from manufacturer.

Guide rails, channels, supports, lattice gates etc. to be minimum 275g/m² galvanized steel.

Fixings to be at least galvanized steel or non ferrous and compatible with other metals used.

Repair/replacement of opening gear and ironmongery as required.

Door operation may be manual or electrical motor driven.

Galvanized steel/aluminium hoods or fascias - lives as for shutter leaves.

Notes

Some fading and chalking of surface coatings may occur within the lives quoted.

A number of manufacturers offer maintenance contracts for the cyclical maintenance and repair of industrial doors.

Key failure modes

Corrosion, scratching/chalking/peeling of surface finish, impact/indentation/distortion, misalignment/ binding/defective operation, failure of operating gear.

Key durability issues

Type/thickness of surface protection, metal thickness/ profile, fixing/support/alignment of guide rails, maintenance frequency, intensity of use, protection against impact.

4.2b

BPG | 4 - Doors, Windows & Joinery

Description

Folding shutter (horizontal) - aluminium

	Description	Maintenance
25	Aluminium shutter leaves to BS 1474/BS EN 485. PVC/Plastisol coating, 200 microns nominal thickness.	Clean every 6 months with non-alkaline detergent (polluted/marine areas every 3 months). Coatings may require redecoration after 10-15 years. Lubrication of joints/bearings in accordance with manufacturer's recommendations, or cyclical maintenance carried out under contract with manufacturer.
20	Aluminium shutter leaves to BS 1474/BS EN 485. Acrylic, PVDF/PVF2, alkyd, polyester or silicone modified polyester coating, 25-50 microns nominal thickness.	Clean every 6 months with non-alkaline detergent (polluted/marine areas every 3 months). Coatings may require redecoration after 10-15 years. Lubrication of joints/bearings in accordance with manufacturer's recommendations, or cyclical maintenance carried out under contract with manufacturer.
20	Anodised aluminium shutter leaves to BS 1474/BS EN 485. Minimum 25 micron coating.	Clean every 6 months with non-alkaline detergent (polluted/marine areas every 3 months). If painted, redecorate every 5 years. Lubrication of joints/bearings in accordance with manufacturer's recommendations, or cyclical maintenance carried out under contract with manufacturer.
10	Anodised aluminium shutter leaves to BS 1474/BS EN 485. Minimum 10 micron coating.	Clean every 6 months with non-alkaline detergent (polluted/marine areas every 3 months). If painted, redecorate every 5 years. Lubrication of joints/bearings in accordance with manufacturer's recommendations, or cyclical maintenance carried out under contract with manufacturer.
5	Mill finished aluminium shutter leaves to BS 1474/BS EN 485.	Clean every 6 months with non-alkaline detergent (polluted/marine areas every 3 months). If painted, redecorate every 5 years. Lubrication of joints/bearings in accordance with manufacturer's recommendations, or cyclical maintenance carried out under contract with manufacturer.
U1	Unclassified, ie aluminium not to relevant BS.	Unclassified.

For Adjustment factors, Assumptions, Notes, Key failure modes and Key durability issues please see overleaf.

4.2c

Industrial doors (external)

LOCATIONS - General

Adjustment factors

Industrial/polluted/marine environment: -5 years.

Guide rails, channels, supports, lattice gates etc. 'factory primed' or less than 275g/m² galvanized steel: life limited to 10 years.

Assumptions

Installation in accordance with manufacturer's instructions.

Leaf thickness to be appropriate for required opening size. Guidance on selection should be obtained from manufacturer.

Guide rails, channels, supports, lattice gates etc. to be minimum 275g/m² galvanized steel.

Fixings to be at least galvanized steel or non ferrous and compatible with other metals used.

Repair/replacement of opening gear and ironmongery as required.

Door operation may be manual or electrical motor driven.

Galvanized steel/aluminium hoods or fascias - lives as for shutter leaves.

Notes

Some fading and chalking of surface coatings may occur within the lives quoted.

A number of manufacturers offer maintenance contracts for the cyclical maintenance and repair of industrial doors.

Key failure modes

Corrosion, scratching/chalking/peeling of surface finish, impact/indentation/distortion, misalignment/binding/defective operation, failure of operating gear.

Key durability issues

Type/thickness of surface protection, metal thickness/profile, fixing/support/alignment of guide rails, maintenance frequency, intensity of use, protection against impact.

4.2c

BPG | **4 - Doors, Windows & Joinery**

General	Description	Maintenance
	Sectional overhead - frameless sections - steel	
20	Post galvanized steel door sections to BS EN 10147, minimum 600g/m² zinc coating weight.	Clean every 6 months with non-alkaline detergent (polluted/marine areas every 3 months). If painted, redecorate every 5 years. Lubrication of joints/bearings in accordance with manufacturer's recommendations, or cyclical maintenance carried out under contract with manufacturer.
15	Post galvanized steel door sections to BS EN 10147, minimum 450g/m² zinc coating weight.	Clean every 6 months with non-alkaline detergent (polluted/marine areas every 3 months). If painted, redecorate every 5 years. Lubrication of joints/bearings in accordance with manufacturer's recommendations, or cyclical maintenance carried out under contract with manufacturer.
15	Pre-galvanized steel door sections to BS 1449:Part 1 or BS EN 10142 minimum 600g/m² zinc coating weight.	Clean every 6 months with non-alkaline detergent (polluted/marine areas every 3 months). If painted, redecorate every 5 years. Lubrication of joints/bearings in accordance with manufacturer's recommendations, or cyclical maintenance carried out under contract with manufacturer.
15	Galvanized steel door sections to BS 1449:Part 1/BS EN 10142 or BS EN 10147, minimum zinc coating 275g/m². PVC/Plastisol coating, 200 microns nominal thickness.	Clean every 6 months with non-alkaline detergent (polluted/marine areas every 3 months). Coatings may require redecoration after 10-15 years. Lubrication of joints/bearings in accordance with manufacturer's recommendations, or cyclical maintenance carried out under contract with manufacturer.
10	Galvanized steel door sections to BS 1449:Part 1/BS EN 10142 or BS EN 10147, minimum zinc coating 275g/m². Acrylic, PVDF/PVF2, alkyd, polyester or silicone modified polyester coating, 25-50 microns nominal thickness.	Clean every 6 months with non-alkaline detergent (polluted/marine areas every 3 months). Coatings may require redecoration after 10-15 years. Lubrication of joints/bearings in accordance with manufacturer's recommendations, or cyclical maintenance carried out under contract with manufacturer.
5	Galvanized steel door sections to BS 1449:Part 1/BS EN 10142 or BS EN 10147, minimum zinc coating 275g/m².	Clean every 6 months with non-alkaline detergent (polluted/marine areas every 3 months). If painted, redecorate every 5 years. Lubrication of joints/bearings in accordance with manufacturer's recommendations, or cyclical maintenance carried out under contract with manufacturer.
5	Mild steel door sections, blast cleaned after any cutting, welding or drilling to BS 7079 'second quality' (equivalent to SA 2.5) and protected with an appropriate Micaceous Iron Oxide, Chlorinated Rubber or similar high performance finish to give a minimum dry film thickness of 250 microns.	Clean every 6 months with non-alkaline detergent (polluted/marine areas every 3 months). Redecorate every 5 years. Lubrication of joints/bearings in accordance with manufacturer's recommendations, or cyclical maintenance carried out under contract with manufacturer.
U1	Unclassified, ie mild steel 'factory primed', steel not to relevant BS.	Unclassified.

4.2d For Adjustment factors, Assumptions, Notes, Key failure modes and Key durability issues please see overleaf.

Industrial doors (external)

LOCATIONS - General

Adjustment factors

Industrial/polluted/marine environment: -5 years.

Guide rails, channels, supports, etc. 'factory primed' or less than 275g/m² galvanized steel: life limited to 10 years.

Fully supported sections (eg rigid foam core): +5 years.

Assumptions

Door sections may be single skin, double skin, insulated double skin with rigid foam core.

Installation in accordance with manufacturer's instructions.

Door leaf/skin thickness to be appropriate for required opening width. Guidance on selection should be obtained from manufacturer.

Guide rails, channels, supports, etc. to be minimum 275g/m² galvanized steel.

Fixings to be at least galvanized steel or non ferrous and compatible with other metals used.

Repair/replacement of opening gear and ironmongery as required.

Door operation may be manual, geared hoist or electrical motor driven.

Galvanized steel/aluminium hoods or fascias - lives as for door sections.

Notes

Some fading and chalking of surface coatings may occur within the lives quoted.

The range of options for protecting steel and aluminium is huge, therefore the options listed above are indicative only.

A number of manufacturers offer maintenance contracts for the cyclical maintenance and repair of industrial doors.

Key failure modes

Corrosion, scratching/chalking/peeling of surface finish, impact/indentation/distortion, misalignment/ binding/defective operation, failure of operating gear.

Key durability issues

Type/thickness of surface protection, thickness/profile of metal skin, support to skin (eg rigid core), fixing/support/alignment of guide rails, maintenance frequency, intensity of use, protection against impact.

LOCATIONS - General

BPG | 4 - Doors, Windows & Joinery

General	Description	Maintenance
Sectional overhead - frameless sections - aluminium		
25	Aluminium door sections to BS 1474/BS EN 485. PVC/Plastisol coating, 200 microns nominal thickness.	Clean every 6 months with non-alkaline detergent (polluted/marine areas every 3 months). If painted, redecorate every 5 years. Coatings may require redecoration after 10-15 years. Lubrication of joints/bearings in accordance with manufacturer's recommendations, or cyclical maintenance carried out under contract with manufacturer.
20	Aluminium door sections to BS 1474/BS EN 485. Acrylic, PVDF/ PVF2, alkyd, polyester or silicone modified polyester coating, 25-50 microns nominal thickness.	Clean every 6 months with non-alkaline detergent (polluted/marine areas every 3 months). Coatings may require redecoration after 10-15 years. Lubrication of joints/bearings in accordance with manufacturer's recommendations, or cyclical maintenance carried out under contract with manufacturer.
20	Anodised aluminium door sections to BS 1474/BS EN 485. Minimum 25 micron coating.	Clean every 6 months with non-alkaline detergent (polluted/marine areas every 3 months). If painted, redecorate every 5 years. Lubrication of joints/bearings in accordance with manufacturer's recommendations, or cyclical maintenance carried out under contract with manufacturer.
10	Anodised aluminium door sections to BS 1474/BS EN 485. Minimum 10 micron coating.	Clean every 6 months with non-alkaline detergent (polluted/marine areas every 3 months). If painted, redecorate every 5 years. Lubrication of joints/bearings in accordance with manufacturer's recommendations, or cyclical maintenance carried out under contract with manufacturer.
5	Mill finished aluminium door sections to BS 1474/BS EN 485.	Clean every 6 months with non-alkaline detergent (polluted/marine areas every 3 months). If painted, redecorate every 5 years. Lubrication of joints/bearings in accordance with manufacturer's recommendations, or cyclical maintenance carried out under contract with manufacturer.
U1	Unclassified, ie aluminium not to relevant BS.	Unclassified.

For Adjustment factors, Assumptions, Notes, Key failure modes and Key durability issues please see overleaf.

4.2e

Industrial doors (external)

SUB TYPES
**Sectional
overhead -
frameless
sections -
aluminium**

LOCATIONS - General

Adjustment factors

Industrial/polluted/marine environment: -5 years.

Guide rails, channels, supports, etc. 'factory primed' or less than 275g/m² galvanized steel: life limited to 10 years.

Fully supported sections (eg rigid foam core): +5 years.

Assumptions

Door sections may be single skin, double skin, insulated double skin with rigid foam core.

Installation in accordance with manufacturer's instructions.

Door leaf/skin thickness to be appropriate for required opening width. Guidance on selection should be obtained from manufacturer.

Guide rails, channels, supports, etc. to be minimum 275g/m² galvanized steel.

Fixings to be at least galvanized steel or non ferrous and compatible with other metals used.

Repair/replacement of opening gear and ironmongery as required.

Door operation may be manual, geared hoist or electrical motor driven.

Galvanized steel/aluminium hoods or fascias - lives as for door sections.

Notes

Some fading and chalking of surface coatings may occur within the lives quoted.

The range of options for protecting steel and aluminium is huge, therefore the options listed above are indicative only.

A number of manufacturers offer maintenance contracts for the cyclical maintenance and repair of industrial doors.

Key failure modes

Corrosion, scratching/chalking/peeling of surface finish, impact/indentation/distortion, misalignment/ binding/defective operation, failure of operating gear.

Key durability issues

Type/thickness of surface protection, thickness/profile of metal skin, support to skin (eg rigid core), fixing/support/alignment of guide rails, maintenance frequency, intensity of use, protection against impact.

LOCATIONS - General

4 - Doors, Windows & Joinery

Sectional overhead - frames

General	Description	Maintenance
30	Aluminium to BS 1474 (extrusions)/BS EN 485 (fabrications/sheet). PVC/Plastisol coating, 200 microns nominal thickness.	Clean every 6 months with non-alkaline detergent (polluted/marine areas every 3 months). Coatings may require redecoration after 10-15 years. Lubrication of joints/bearings in accordance with manufacturer's recommendations, or cyclical maintenance carried out under contract with manufacturer.
25	Aluminium to BS 1474 (extrusions)/BS EN 485 (fabrications/sheet). Acrylic, PVDF/PVF2, alkyd, polyester or silicone modified polyester coating, 25-50 microns nominal thickness.	Clean every 6 months with non-alkaline detergent (polluted/marine areas every 3 months). Coatings may require redecoration after 10-15 years. Lubrication of joints/bearings in accordance with manufacturer's recommendations, or cyclical maintenance carried out under contract with manufacturer.
25	Aluminium to BS 1474 (extrusions)/BS EN 485 (fabrications/sheet). Anodized to BS 1615, minimum 25 micron coating.	Clean every 6 months with non-alkaline detergent (polluted/marine areas every 3 months). If painted, redecorate every 5 years. Lubrication of joints/bearings in accordance with manufacturer's recommendations, or cyclical maintenance carried out under contract with manufacturer.
15	Aluminium to BS 1474 (extrusions)/BS EN 485 (fabrications/sheet). Anodized to BS 1615, minimum 10 micron coating.	Clean every 6 months with non-alkaline detergent (polluted/marine areas every 3 months). If painted, redecorate every 5 years. Lubrication of joints/bearings in accordance with manufacturer's recommendations, or cyclical maintenance carried out under contract with manufacturer.
10	Aluminium to BS 1474 (extrusions)/BS EN 485 (fabrications/sheet). Mill finished.	Clean every 6 months with non-alkaline detergent (polluted/marine areas every 3 months). If painted, redecorate every 5 years. Lubrication of joints/bearings in accordance with manufacturer's recommendations, or cyclical maintenance carried out under contract with manufacturer.
U1	Unclassified, ie aluminium, not to relevant BS.	Unclassified.

For Adjustment factors, Assumptions, Notes, Key failure modes and Key durability issues please see overleaf.

4.2f

Industrial doors (external)

Adjustment factors

Industrial/polluted/marine environment: -5 years.

Guide rails, channels, supports, etc. 'factory primed' or less than 275g/m^2 galvanized steel: life limited to 10 years.

Assumptions

Installation in accordance with manufacturer's instructions.

Guide rails, channels, supports, etc. to be minimum 275g/m^2 galvanized steel.

Fixings to be at least galvanized steel or non ferrous and compatible with other metals used.

Repair/replacement of opening gear and ironmongery as required.

Door operation may be manual, geared hoist or electrical motor driven.

Notes

Some fading and chalking of surface coatings may occur within the lives quoted.

A number of manufacturers offer maintenance contracts for the cyclical maintenance and repair of industrial doors.

Key failure modes

Corrosion, scratching/chalking/peeling of surface finish, impact, misalignment/binding/defective operation, failure of operating gear.

Key durability issues

Type/thickness of surface protection, fixing/support/alignment of guide rails, maintenance frequency, intensity of use, protection against impact.

BPG | 4 - Doors, Windows & Joinery

Maintenance

Clean every 6 months with non-alkaline detergent (polluted/marine areas every 3 months). If painted, redecorate every 5 years. Lubrication of joints/bearings in accordance with manufacturer's recommendations, or cyclical maintenance carried out under contract with manufacturer.

Clean every 6 months with non-alkaline detergent (polluted/marine areas every 3 months). If painted, redecorate every 5 years. Lubrication of joints/bearings in accordance with manufacturer's recommendations, or cyclical maintenance carried out under contract with manufacturer.

Clean every 6 months with non-alkaline detergent (polluted/marine areas every 3 months). If painted, redecorate every 5 years. Lubrication of joints/bearings in accordance with manufacturer's recommendations, or cyclical maintenance carried out under contract with manufacturer.

Clean every 6 months with non-alkaline detergent (polluted/marine areas every 3 months). Coatings may require redecoration after 10-15 years. Lubrication of joints/bearings in accordance with manufacturer's recommendations, or cyclical maintenance carried out under contract with manufacturer.

Clean every 6 months with non-alkaline detergent (polluted/marine areas every 3 months). Coatings may require redecoration after 10-15 years. Lubrication of joints/bearings in accordance with manufacturer's recommendations, or cyclical maintenance carried out under contract with manufacturer.

Clean every 6 months with non-alkaline detergent (polluted/marine areas every 3 months). If painted, redecorate every 5 years. Lubrication of joints/bearings in accordance with manufacturer's recommendations, or cyclical maintenance carried out under contract with manufacturer.

Clean every 6 months with non-alkaline detergent (polluted/marine areas every 3 months). Redecorate every 5 years. Lubrication of joints/bearings in accordance with manufacturer's recommendations, or cyclical maintenance carried out under contract with manufacturer.

Unclassified.

Description

Sectional overhead - metal infill panels - steel

20 Post galvanized steel panels to BS EN 10147, minimum 600g/m^2 zinc coating weight.

15 Post galvanized steel panels to BS EN 10147, minimum 450g/m^2 zinc coating weight.

15 Pre-galvanized steel panels to BS 1449:Part 1 or BS EN 10142 minimum 600g/m^2 zinc coating weight.

15 Galvanized steel panels to BS 1449:Part 1/BS EN 10142 or BS EN 10147, minimum zinc coating 275g/m^2. PVC/Plastisol coating, 200 microns nominal thickness.

10 Galvanized steel panels to BS 1449:Part 1/BS EN 10142 or BS EN 10147, minimum zinc coating 275g/m^2. Acrylic, PVDF/PVF2, alkyd, polyester or silicone modified polyester coating, 25-50 microns nominal thickness.

5 Galvanized steel panels to BS 1449:Part 1/BS EN 10142 or BS EN 10147, minimum zinc coating 275g/m^2.

5 Mild steel panels, blast cleaned after any cutting, welding or drilling to BS 7079 'second quality' (equivalent to SA 2.5) and protected with an appropriate Micaceous Iron Oxide, Chlorinated Rubber or similar high performance finish to give a minimum dry film thickness of 250 microns.

U1 Unclassified, ie mild steel 'factory primed', steel not to relevant BS.

4.2g | For Adjustment factors, Assumptions, Notes, Key failure modes and Key durability issues please see overleaf.

Industrial doors (external)

LOCATIONS - General

Adjustment factors

Industrial/polluted/marine environment: -5 years.

Fully supported sections (eg rigid foam core): +5 years.

Assumptions

Infill panels may be single skin, double skin, insulated double skin with rigid foam core.

Installation in accordance with manufacturer's instructions.

Fixings to be at least galvanized steel or non ferrous and compatible with other metals used.

Galvanized steel/aluminium hoods or fascias - lives as for infill panels.

Notes

Some fading and chalking of surface coatings may occur within the lives quoted.

See previous page for frame sections.

For glazed infill panels, see balustrades, P.8.4a.

Key failure modes

Corrosion, scratching/chalking/peeling of surface finish, impact/indentation/distortion.

Key durability issues

Type/thickness of surface protection, thickness/profile of metal skin, support to skin (eg rigid core), fixing of panels in frame, maintenance frequency, protection against impact.

Industrial doors (external)

BPG	4 - Doors, Windows & Joinery	
General	**Description**	**Maintenance**
	Sectional overhead - metal infill panels - aluminium	
25	Aluminium panels to BS 1474/BS EN 485. PVC/Plastisol coating, 200 microns nominal thickness.	Clean every 6 months with non-alkaline detergent (polluted/ marine areas every 3 months). Coatings may require redecoration after 10-15 years. Lubrication of joints/bearings in accordance with manufacturer's recommendations, or cyclical maintenance carried out under contract with manufacturer.
20	Aluminium panels to BS 1474/BS EN 485. Acrylic, PVDF/ PVF2, alkyd, polyester or silicone modified polyester coating, 25-50 microns nominal thickness.	Clean every 6 months with non-alkaline detergent (polluted/ marine areas every 3 months). Coatings may require redecoration after 10-15 years. Lubrication of joints/bearings in accordance with manufacturer's recommendations, or cyclical maintenance carried out under contract with manufacturer.
20	Anodised aluminium panels to BS 1474/BS EN 485. Minimum 25 micron coating.	Clean every 6 months with non-alkaline detergent (polluted/ marine areas every 3 months). If painted, redecorate every 5 years. Lubrication of joints/bearings in accordance with manufacturer's recommendations, or cyclical maintenance carried out under contract with manufacturer.
10	Anodised aluminium panels to BS 1474/BS EN 485. Minimum 10 micron coating.	Clean every 6 months with non-alkaline detergent (polluted/ marine areas every 3 months). If painted, redecorate every 5 years. Lubrication of joints/bearings in accordance with manufacturer's recommendations, or cyclical maintenance carried out under contract with manufacturer.
5	Mill finished aluminium panels to BS 1474/BS EN 485.	Clean every 6 months with non-alkaline detergent (polluted/ marine areas every 3 months). If painted, redecorate every 5 years. Lubrication of joints/bearings in accordance with manufacturer's recommendations, or cyclical maintenance carried out under contract with manufacturer.
U1	Unclassified, ie aluminium not to relevant BS.	Unclassified.

For Adjustment factors, Assumptions, Notes, Key failure modes and Key durability issues please see overleaf.

4.2h

Industrial doors (external)

LOCATIONS - General

BPG 4 - Doors, Windows & Joinery

Adjustment factors

Industrial/polluted/marine environment: -5 years.

Fully supported sections (eg rigid foam core): +5 years.

Assumptions

Infill panels may be single skin, double skin, insulated double skin with rigid foam core.

Installation in accordance with manufacturer's instructions.

Fixings to be at least galvanized steel or non ferrous and compatible with other metals used.

Galvanized steel/aluminium hoods or fascias - lives as for infill panels.

Notes

Some fading and chalking of surface coatings may occur within the lives quoted.

See previous page for frame sections.

For glazed infill panels, see balustrades, P.8.4a.

Key failure modes

Corrosion, scratching/chalking/peeling of surface finish, impact/indentation/distortion.

Key durability issues

Type/thickness of surface protection, thickness/profile of metal skin, support to skin (eg rigid core), fixing of panels in frame, maintenance frequency, protection against impact.

4.2h

Industrial doors (external)

LOCATIONS - General

4 - Doors, Windows & Joinery

General	Description	Maintenance
Folding/sliding sectional - steel		
20	Post galvanized steel door leaves to BS EN 10147, minimum 600g/m² zinc coating weight.	Clean every 6 months with non-alkaline detergent (polluted/marine areas every 3 months). If painted, redecorate every 5 years. Lubrication of joints/bearings in accordance with manufacturer's recommendations, or cyclical maintenance carried out under contract with manufacturer.
15	Post-galvanized steel door leaves to BS EN 10147, minimum 450g/m² zinc coating weight.	Clean every 6 months with non-alkaline detergent (polluted/marine areas every 3 months). If painted, redecorate every 5 years. Lubrication of joints/bearings in accordance with manufacturer's recommendations, or cyclical maintenance carried out under contract with manufacturer.
15	Pre-galvanized steel door leaves to BS EN 10142 minimum 600g/m² zinc coating weight.	Clean every 6 months with non-alkaline detergent (polluted/marine areas every 3 months). If painted, redecorate every 5 years. Lubrication of joints/bearings in accordance with manufacturer's recommendations, or cyclical maintenance carried out under contract with manufacturer.
15	Galvanized steel door leaves to BS 1449:Part 1/BS EN 10142 or BS EN 10147, minimum zinc coating 275g/m². PVC/Plastisol coating, 200 microns nominal thickness.	Clean every 6 months with non-alkaline detergent (polluted/marine areas every 3 months).Coatings may require redecoration after 10-15 years. Lubrication of joints/bearings in accordance with manufacturer's recommendations, or cyclical maintenance carried out under contract with manufacturer.
10	Galvanized steel door leaves to BS 1449:Part 1/BS EN 10142 or BS EN 10147, minimum zinc coating 275g/m². Acrylic, PVDF/PVF2, alkyd, polyester or silicone modified polyester coating, 25-50 microns nominal thickness.	Clean every 6 months with non-alkaline detergent (polluted/marine areas every 3 months). Coatings may require redecoration after 10-15 years. Lubrication of joints/bearings in accordance with manufacturer's recommendations, or cyclical maintenance carried out under contract with manufacturer.
5	Galvanized steel door leaves to BS 1449:Part 1/BS EN 10142 or BS EN 10147, minimum zinc coating 275g/m².	Clean every 6 months with non-alkaline detergent (polluted/marine areas every 3 months). If painted, redecorate every 5 years. Lubrication of joints/bearings in accordance with manufacturer's recommendations, or cyclical maintenance carried out under contract with manufacturer.
5	Mild steel door leaves, blast cleaned after any cutting, welding or drilling to BS 7079 'second quality' (equivalent to SA 2.5) and protected with an appropriate Micaceous Iron Oxide, Chlorinated Rubber or similar high performance finish to give a minimum dry film thickness of 250 microns.	Clean every 6 months with non-alkaline detergent (polluted/marine areas every 3 months). Redecorate every 5 years. Lubrication of joints/bearings in accordance with manufacturer's recommendations, or cyclical maintenance carried out under contract with manufacturer.
U1	Unclassified, ie mild steel 'factory primed', steel not to relevant BS.	Unclassified

4.2j For Adjustment factors, Assumptions, Notes, Key failure modes and Key durability issues please see overleaf.

Industrial doors (external)

LOCATIONS - General

Adjustment factors

Industrial/polluted/marine environment: -5 years.

Guide rails, channels, supports, etc. 'factory primed' or less than 275g/m^2 galvanized steel: life limited to 10 years.

Fully supported sections (eg rigid foam core): +5 years.

Assumptions

Door sections may be single skin, double skin, insulated double skin with rigid foam core.

Installation in accordance with manufacturer's instructions.

Guide rails, channels, supports, etc. to be minimum 275g/m^2 galvanized steel.

Fixings to be at least galvanized steel or non ferrous and compatible with other metals used.

Repair/replacement of opening gear and ironmongery as required.

Door operation may be manual or electrical motor driven.

Galvanized steel/aluminium hoods or fascias - lives as for door leaves.

Notes

Some fading and chalking of surface coatings may occur within the lives quoted.

A number of manufacturers offer maintenance contracts for the cyclical maintenance and repair of industrial doors.

Key failure modes

Corrosion, scratching/chalking/peeling of surface finish, impact/indentation/distortion, misalignment/ binding/defective operation, failure of operating gear.

Key durability issues

Type/thickness of surface protection, thickness/profile of metal skin, support to skin (eg rigid core), fixing/support/ alignment of guide rails, maintenance frequency, intensity of use, protection against impact.

Industrial doors (external)

LOCATIONS - General

BPG | 4 - Doors, Windows & Joinery

General	**Description**		**Maintenance**
	Folding/sliding sectional - aluminium		
25	Aluminium door leaves to BS 1474/BS EN 485. PVC/Plastisol coating, 200 microns nominal thickness.		Clean every 6 months with non-alkaline detergent (polluted/ marine areas every 3 months). Coatings may require redecoration after 10-15 years. Lubrication of joints/bearings in accordance with manufacturer's recommendations, or cyclical maintenance carried out under contract with manufacturer.
20	Aluminium door leaves to BS 1474/BS EN 485. Acrylic, PVDF/ PVF2, alkyd, polyester or silicone modified polyester coating, 25-50 microns nominal thickness.		Clean every 6 months with non-alkaline detergent (polluted/ marine areas every 3 months).Coatings may require redecoration after 10-15 years. Lubrication of joints/bearings in accordance with manufacturer's recommendations, or cyclical maintenance carried out under contract with manufacturer.
20	Anodised aluminium door leaves to BS 1474/BS EN 485. Minimum 25 micron coating.		Clean every 6 months with non-alkaline detergent (polluted/ marine areas every 3 months). If painted, redecorate every 5 years. Lubrication of joints/bearings in accordance with manufacturer's recommendations, or cyclical maintenance carried out under contract with manufacturer.
10	Anodised aluminium door leaves to BS 1474/BS EN 485. Minimum 10 micron coating.		Clean every 6 months with non-alkaline detergent (polluted/ marine areas every 3 months). If painted, redecorate every 5 years. Lubrication of joints/bearings in accordance with manufacturer's recommendations, or cyclical maintenance carried out under contract with manufacturer.
5	Mill finished aluminium door leaves to BS 1474/BS EN 485.		Clean every 6 months with non-alkaline detergent (polluted/ marine areas every 3 months). If painted, redecorate every 5 years. Lubrication of joints/bearings in accordance with manufacturer's recommendations, or cyclical maintenance carried out under contract with manufacturer.
U1	Unclassified, ie aluminium not to relevant BS.		Unclassified

For Adjustment factors, Assumptions, Notes, Key failure modes and Key durability issues please see overleaf.

4.2k

Industrial doors (external)

LOCATIONS - General

Adjustment factors

Industrial/polluted/marine environment: -5 years.

Guide rails, channels, supports, etc. 'factory primed' or less than 275g/m² galvanized steel: life limited to 10 years.

Fully supported sections (eg rigid foam core): +5 years.

Assumptions

Door sections may be single skin, double skin, insulated double skin with rigid foam core.

Installation in accordance with manufacturer's instructions.

Guide rails, channels, supports, etc. to be minimum 275g/m² galvanized steel.

Fixings to be at least galvanized steel or non ferrous and compatible with other metals used.

Repair/replacement of opening gear and ironmongery as required.

Door operation may be manual or electrical motor driven.

Galvanized steel/aluminium hoods or fascias - lives as for door leaves.

Notes

Some fading and chalking of surface coatings may occur within the lives quoted.

A number of manufacturers offer maintenance contracts for the cyclical maintenance and repair of industrial doors.

Key failure modes

Corrosion, scratching/chalking/peeling of surface finish, impact/indentation/distortion, misalignment/ binding/defective operation, failure of operating gear.

Key durability issues

Type/thickness of surface protection, thickness/profile of metal skin, support to skin (eg rigid core), fixing/support/ alignment of guide rails, maintenance frequency, intensity of use, protection against impact.

Industrial doors (external)

LOCATIONS - General

| 4 - Doors, Windows & Joinery

Description

Fast action traffic doors

10   PVC coated nylon fabric curtain, minimum weight 650g/m². Curtain stiffened with 19mm hollow galvanized steel tubular stiffeners.

10   Polyester fabric curtain (2-ply), minimum 1.4mm thick. Urethane impregnated both sides.

10   Clear PVC curtain, minimum 4mm thick.

U1   Unclassified, ie nylon/polyester fabric/PVC curtain less than specification above.

Maintenance

Cleaning in accordance with manufacturer's recommendations. Regular maintenance of electrical motor/cyclical maintenance carried out under contract with manufacturer.

Cleaning in accordance with manufacturer's recommendations. Regular maintenance of electrical motor/cyclical maintenance carried out under contract with manufacturer.

Cleaning in accordance with manufacturer's recommendations. Regular maintenance of electrical motor/cyclical maintenance carried out under contract with manufacturer.

Unclassified.

Adjustment factors

Guide rails, channels, supports, etc. 'factory primed' or less than 275g/m² galvanized steel: -5 years.

Assumptions

Installation in accordance with manufacturer's instructions.

Guide rails, channels, supports, etc. to be minimum 275g/m² galvanized steel.

Fixings to be at least galvanized steel or non ferrous and compatible with other metals used.

Repair/replacement of opening gear and ironmongery as required.

Door operation is electrical motor driven.

Notes

Fast action traffic doors are vertically opening, flexible curtains which open in two directions from the middle, and are power operated. 'Push open' and other flexible curtains are excluded from this section.

A number of manufacturers offer maintenance contracts for the cyclical maintenance and repair of industrial doors.

Key failure modes

Tearing/puncturing of curtain, misalignment/binding/defective operation, failure of motor.

Key durability issues

Material thickness, stiffness, intensity of use, fixing/support/alignment of guide rails, maintenance frequency.

Door ironmongery

Scope

This section provides data on ironmongery for internal and external door leafs for use in non-domestic building types. It includes hinges, levers and latches, locking mechanisms, emergency exit devices and other door furniture. Door closers are covered in Section 4.5 of this manual.

The following component sub-types are included within this section:

	Page
Internal door hinges	4.3
External door hinges	4.3
Push/pull handles/plates	4.3a
Lever & knob furniture	4.3b
Emergency exit devices	4.3c
Locks/latches	4.3d
Mechanical push button locks	4.3e

It should be noted that there is an enormous range of different types and compositions of door ironmongery available to UK specifiers. This section aims to cover the most commonly used materials and types

Standards cited

BS 1344: Part 2:1975 (1994)	Methods of testing vitreous enamel finishes.
	Resistence to citric acid at room temperature.
BS 1615:1987 (1994)	Method for specifying anodic oxidation coatings on aluminium and its alloys.
BS 3621:1980	Specification for thief resistence locks.
BS 4951:1973(1995)	Specification for builders hardware: lock and latch furniture (doors).
BS 5872:1980(1995)	Specification for locks and latches for doors in buildings.

BS 6496:1984(1991)	Specification for powder organic coatings for application and stoving to aluminium alloy extrusions, sheet and preformed sections, for external architectural purposes, and for the finish on aluminium alloy extrusions, sheet and preformed sections coated with powder organic coatings.
BS 7352:1990	Specification for strength and durability performance of metal hinges for side hanging applications and dimensional requirements for template drilled hinges.
BS 7479:1991	Method for salt spray corrosion tests in artificial atmospheres.
BS PD 6484:1979 (1990)	Commentary on corrosion on bimetallic contacts and it's alleviation.
BS EN 179:1998	Building hardware. Emergency exit devices operated by a lever handle or push pad. Requirements and test methods.
BS EN 1125:1997	Building hardware - panic exit devices operated by a horizontal bar. requirements and test methods.
BS EN 1670:1998	Building hardware - corrosion resistance - requirements and test methods.

Other references/information sources

Association of Builders' Hardware Manufacturers Codes of Practice.

Guild of Architectural Ironmongers Guidance Notes.

PSA Method of Building Product Selection Guides.

Door ironmongery

SUB TYPES
Internal door hinges
External door hinges

BPG **4 - Doors, Windows & Joinery**

General	Description	Maintenance
	Internal door hinges	
30	Hinges to Classes 3 or 4 of BS 7352 (ie suitable for light to medium duty as defined in the BSI for doors not generally used by the public or by people carrying or propelling heavy items, located in normal, heated, indoor environments. Corrosion resistance to appropriate BS EN 1670 class (see assumptions).	Periodic lubrication (ie when hinges become noisy). If site painted, redecorate every 5 years.
25	Hinges to Classes 8 or 9 of BS 7352 (ie suitable for severe duty use on averagely trafficked routes) where doors are frequently subjected to violent use and impacts, and Class 5 (ie suitable for heavy duty use on averagely trafficked routes) where users have little incentive to take care. Corrosion resistance to appropriate BS EN 1670 class (see assumptions).	Periodic lubrication (ie when hinges become noisy). If site painted, redecorate every 5 years.
20	Hinges to Classes 6 or 7 of BS 7352 (ie suitable for heavy duty use on heavily trafficked routes) where users have little incentive to take care. Corrosion resistance to appropriate BS EN 1670 class (see assumptions).	Periodic lubrication (ie when hinges become noisy). If site painted, redecorate every 5 years.
5	Hinges not classified to BS 7352, or corrosion resistance not to appropriate BS EN 1670 class.	Periodic lubrication (ie when hinges become noisy). If site painted, redecorate every 5 years.
	External door hinges	
25	Hinges to Classes 3 or 4 of BS 7352 (ie suitable for light to medium duty as defined in the BSI for doors not generally used by the public or by people carrying or propelling heavy items. Corrosion resistance to appropriate BS EN 1670 class (see assumptions).	Periodic lubrication (ie when hinges become noisy). If site painted, redecorate every 5 years.
20	Hinges to Class 5 of BS 7352 (ie suitable for heavy duty use on averagely trafficked routes) where users have little incentive to take care, and to Classes 8 or 9 (ie suitable for severe duty use on averagely trafficked routes) where doors are frequently subjected to violent use and impacts. Corrosion resistance to appropriate BS EN 1670 class (see assumptions).	Periodic lubrication (ie when hinges become noisy). If site painted, redecorate every 5 years.
15	Hinges to Classes 6 or 7 of BS 7352 (ie suitable for heavy duty use on heavily trafficked routes) where users have little incentive to take care. Corrosion resistance to appropriate BS EN 1670 class (see assumptions).	Periodic lubrication (ie when hinges become noisy). If site painted, redecorate every 5 years.
U1	Unclassified, ie hinges not classified to BS 7352, or corrosion resistance not to appropriate BS EN 1670 class.	Unclassified.

For Adjustment factors, Assumptions, Notes, Key failure modes and Key durability issues please see overleaf.

4.3

Door ironmongery

LOCATIONS - General

SUB TYPES
Internal door hinges
External door hinges

BPG   4 - Doors, Windows & Joinery

Adjustment factors

Periodic site applied protective treatment in lieu of factory applied: -5 years (external hinges only).

Corrosion resistance less than Class 3 to BS EN 1670: -5 years.

Assumptions

BS 7352 defines 9 classes of hinges according to strength and allowable wear. The appropriate hinge class must be selected for the mass of the door leaf.

The BS 7352 strength classes are based on the assumption that 3 hinges are provided for each door leaf. Where two or four hinges are provided, the door leaf mass should be factored accordingly.

Where the door is fitted with an overhead door closer, the hinge should be minimum Class 5 'heavy duty' to BS 7352.

Hinges to be tested to comply with appropriate corrosion resistance class to BS EN 1670, ie Class 1: dry internal environment; Class 2: occasionally damp interiors, areas subject to condensation; Class 3: interiors, areas that are often wet or subject to slight pollution (ie most external environments), particularly damp interiors; Class 4: very polluted environments, eg combinations of industrial and coastal pollution.

Unplated steel fixings should only be used where the hinge is to be painted in-situ. Otherwise, fixings should be protectively treated (eg zinc plated) or of inherently corrosion resistant materials such as stainless steel, brass or aluminium. Fixings to be compatible with the hinge material, ie to prevent galvanic corrosion. For further detailed guidance, see BS PD 6484.

Notes

Hinges to BS 7352 must be marked with the manufacturer's identity, the BS number and hinge class.

Corrosion resistance beyond BS EN 1670 Class 4 may be required in highly polluted environments.

Key failure modes

Corrosion, wear, misalignment, distortion.

Key durability issues

Base metal, surface protection, frequency of use, degree of care in use, size/specification of hinge, maintenance frequency.

4.3

TYPE

Door ironmongery ___

LOCATIONS - Internal, External

SUB TYPES
Push/pull
handles/plates

BPG | 4 - Doors, Windows & Joinery

Internal	External	Description	Maintenance
		Push/pull handles/plates	
35+	35+	Grade 304/316 austenitic stainless steel loop handles with austenitic stainless steel fixings. Corrosion resistance to appropriate BS EN 1670 class (see assumptions).	Periodic cleaning in accordance with manufacturer's recommendations.
35	30	One-piece tubular brass loop handles. Corrosion resistance to appropriate BS EN 1670 class (see assumptions).	Periodic cleaning in accordance with manufacturer's recommendations.
35	30	Tubular brass straight handles with separate cast brass ends. Corrosion resistance to appropriate BS EN 1670 class (see assumptions).	Periodic cleaning in accordance with manufacturer's recommendations.
25	25	One-piece tubular aluminium alloy loop handles. Anodized to BS 1615, minimum 20 micron coating thickness. Corrosion resistance to appropriate BS EN 1670 class (see assumptions).	Periodic cleaning in accordance with manufacturer's recommendations.
25	25	Tubular aluminium alloy straight handles with separate cast LM5 alloy ends. Anodized to BS 1615, minimum 20 micron coating thickness. Corrosion resistance to appropriate BS EN 1670 class (see assumptions).	Periodic cleaning in accordance with manufacturer's recommendations.
25	15	One-piece tubular aluminium alloy loop handles. Anodized to BS 1615, minimum 15 micron coating thickness. Corrosion resistance to appropriate BS EN 1670 class (see assumptions).	Periodic cleaning in accordance with manufacturer's recommendations.
25	15	Tubular aluminium alloy straight handles with separate cast LM5 alloy ends. Anodized to BS 1615, minimum 15 micron coating thickness. Corrosion resistance to appropriate BS EN 1670 class (see assumptions).	Periodic cleaning in accordance with manufacturer's recommendations.
20	15	One-piece tubular aluminium alloy loop handles. Polyester powder coating to BS 6496, minimum 25 micron coating thickness. Corrosion resistance to appropriate BS EN 1670 class (see assumptions).	Periodic cleaning in accordance with manufacturer's recommendations.
20	15	Tubular aluminium alloy straight handles with separate cast LM5 alloy ends. Polyester powder coating to BS 6496, minimum 25 micron coating thickness. Corrosion resistance to appropriate BS EN 1670 class (see assumptions).	Periodic cleaning in accordance with manufacturer's recommendations.
20	15	One-piece tubular mild steel loop handles with vitreous enamel coating to Class B of BS 1344:Part 2. Corrosion resistance to appropriate BS EN 1670 class (see assumptions).	Periodic cleaning in accordance with manufacturer's recommendations.
10	5	Brass plated steel. Corrosion resistance to appropriate BS EN 1670 class (see assumptions).	Periodic cleaning in accordance with manufacturer's recommendations.
5	U1	Metal handles, surface protection less than those stated above. Corrosion resistance to appropriate BS EN 1670 class (see assumptions).	Periodic cleaning in accordance with manufacturer's recommendations.

For Adjustment factors, Assumptions, Notes, Key failure modes and Key durability issues please see overleaf.

Door ironmongery

LOCATIONS - Internal, External

BPG 4 - Doors, Windows & Joinery

Adjustment factors

Internal doors in heavily trafficked (eg main corridor) routes: -5 years.

Corrosion resistance less than Class 3 to BS EN 1670: -5 years.

Assumptions

Push/pull handles are through bolted, not face-fixed.

External doors are assumed to be main entrance doors with correspondingly heavy usage.

Ironmongery to be tested to comply with appropriate corrosion resistance class to BS EN 1670, ie Class 1: dry internal environment; Class 2: occasionally damp interiors, areas subject to condensation; Class 3: environments that are often wet or subject to slight pollution (ie most external environments), particularly damp interiors; Class 4: very polluted environments, eg combinations of industrial and coastal pollution.

Unplated steel fixings should only be used where the ironmongery is to be painted in-situ. Otherwise, fixings should be protectively treated (eg zinc plated) or of inherently corrosion resistant materials such as stainless steel, brass or aluminium. Fixings to be compatible with the ironmongery material, ie to prevent galvanic corrosion. For further detailed guidance, see BS PD 6484.

Notes

A number of European Standards for ironmongery are currently under development which will include a standard coding system for performance characteristics such as durability and corrosion resistance. This page will be amended to take account of the new standards once they are introduced.

Corrosion resistance beyond BS EN 1670 Class 4 may be required in highly polluted environments.

Key failure modes

Corrosion, chalking/peeling/loss of surface coating.

Key durability issues

Base metal, surface protection, frequency of use, degree of care in use, maintenance frequency.

Door ironmongery

LOCATIONS - Internal, External

4 - Doors, Windows & Joinery

Internal	External	Description	Maintenance
		Lever & knob furniture (for use with locks & latches)	
35+	35+	Grade 304/316 austenitic stainless steel. Corrosion resistance to appropriate BS EN 1670 class (see assumptions).	Periodic cleaning & lubrication in accordance with manufacturer's recommendations.
35	30	Brass. Corrosion resistance to appropriate BS EN 1670 class (see assumptions).	Periodic cleaning & lubrication in accordance with manufacturer's recommendations.
25	25	Aluminium alloy, anodised to BS 1615, minimum 20 micron coating thickness. Corrosion resistance to appropriate BS EN 1670 class (see assumptions).	Periodic cleaning & lubrication in accordance with manufacturer's recommendations.
25	15	Aluminium alloy, anodised to BS 1615, minimum 15 micron coating thickness. Corrosion resistance to appropriate BS EN 1670 class (see assumptions).	Periodic cleaning & lubrication in accordance with manufacturer's recommendations.
20	15	Aluminium alloy, polyester powder coated to BS 6496, minimum 25 micron coating thickness. Corrosion resistance to appropriate BS EN 1670 class (see assumptions).	Periodic cleaning & lubrication in accordance with manufacturer's recommendations.
20	15	Mild steel with vitreous enamel coating to Class B of BS 1344:Part 2. Corrosion resistance to appropriate BS EN 1670 class (see assumptions).	Periodic cleaning & lubrication in accordance with manufacturer's recommendations.
15	10	Diecast zinc, nickel and chromium plated. Corrosion resistance to appropriate BS EN 1670 class (see assumptions).	Periodic cleaning & lubrication in accordance with manufacturer's recommendations.
10	5	Brass plated steel. Corrosion resistance to appropriate BS EN 1670 class (see assumptions).	Periodic cleaning & lubrication in accordance with manufacturer's recommendations.
10	U1	Diecast zinc, nickel plated. Corrosion resistance to appropriate BS EN 1670 class (see assumptions).	Periodic cleaning & lubrication in accordance with manufacturer's recommendations.

For Adjustment factors, Assumptions, Notes, Key failure modes and Key durability issues please see overleaf.

4.3b

LOCATIONS - Internal, External

Notes

There is a vast number of materials and surface finishes available for door furniture. The above descriptions are indicative only.

A number of European Standards for ironmongery are currently under development which will include a standard coding system for performance characteristics such as durability and corrosion resistance. This page will be amended to take account of the new standards once they are introduced.

Corrosion resistance beyond BS EN 1670 Class 4 may be required in highly polluted environments.

Door furniture to BS 4951 is marked with the manufacturer's identity and the BS number.

Key failure modes

Corrosion, chalking/peeling/loss of surface coating, binding/failure of mechanism.

Key durability issues

Base metal, surface protection, frequency of use, degree of care in use, maintenance frequency.

Adjustment factors

Corrosion resistance less than Class 3 to BS EN 1670: -5 years.

Internal doors in heavily trafficked (eg main corridor) routes: -5 years.

Assumptions

External doors are assumed to be main entrance doors with correspondingly heavy usage.

Door furniture to BS 4951, Category 1 (heavy duty, public use).

Ironmongery to be tested to comply with appropriate corrosion resistance class to BS EN 1670, ie Class 1: dry internal environment; Class 2: occasionally damp interiors, areas subject to condensation; Class 3: environments that are often wet or subject to slight pollution (ie most external environments), particularly damp interiors; Class 4: very polluted environments, eg combinations of industrial and coastal pollution.

Fixings should be protectively treated (eg zinc plated) or of inherently corrosion resistant materials such as stainless steel, brass or aluminium. Fixings to be compatible with the lever/knob material, ie to prevent galvanic corrosion. For further detailed guidance, see BS PD 6484.

4.3b

Door ironmongery

LOCATIONS - Internal

BPG **4 - Doors, Windows & Joinery**

Description

Internal		Maintenance
Emergency exit devices		
35	Emergency exit devices operated by lever handle or push pad, complying with BS EN 179, durability Grade 7 and with corrosion resistance to at least Grade 3 of BS EN 1670.	Periodic lubrication in accordance with manufacturer's instructions.
25	Emergency exit devices operated by lever handle or push pad, complying with BS EN 179, durability Grade 6 and with corrosion resistance to at least Grade 3 of BS EN 1670.	Periodic lubrication in accordance with manufacturer's instructions.
25	Horizontal panic bolts complying with BS EN 1125, durability Grade 7 and with corrosion resistance to at least Grade 3 of BS EN 1670.	Periodic lubrication in accordance with manufacturer's instructions.
20	Horizontal panic bolts complying with BS EN 1125, durability Grade 6 and with corrosion resistance to at least Grade 3 of BS EN 1670.	Periodic lubrication in accordance with manufacturer's instructions.
U1	Unclassified, ie emergency exit devices not complying with BS EN 179 or BS EN 1125, or corrosion resistance less than Grade 3 of BS EN 1670.	Unclassified.

Adjustment factors

None.

Assumptions

Emergency devices to BS EN 179 and BS EN 1125 are intended for use on hinged or pivoted door leaves not exceeding 200 kg in mass, 2500 mm in height and 1300 mm in width.

The BS EN 179 and BS EN 1125 durability tests assume a high frequency of use by the public and others with little incentive to exercise care.

Corrosion resistance Grade 3 to BS EN 1670 is broadly equivalent to 96 hours' neutral salt spray testing to BS 5466.

Unplated steel fixings should only be used where the ironmongery is to be painted in-situ. Otherwise, fixings should be protectively treated (eg zinc plated) or of inherently corrosion resistant materials such as stainless steel, brass or aluminium. Fixings to be compatible with the exit device material, ie to prevent galvanic corrosion. For further detailed guidance, see BS PD 6484.

Notes

Emergency exit devices to BS EN 179 and BS EN 1125 are marked with the manufacturer's identity, BS No. and classification (ie for strength and durability), and the month and year of assembly.

BS EN 179 and BS EN 1125 recommend monthly inspection and operation to ensure that all components are in a satisfactory working order.

Key failure modes

Corrosion, wear, misalignment, distortion.

Key durability issues

Base metal, surface protection, frequency of use, degree of care in use, maintenance frequency.

4.3c

Door ironmongery

LOCATIONS - General

General	Description	Maintenance
	Locks/latches	
20	Mortice locks/latches complying with ANSI A156.5 Grade 1. All exposed parts of austenitic stainless steel, bronze or brass and lock case (within door) of galvanized steel. Latch and deadbolt tongues of austenitic stainless steel, brass or brass with hardened steel rollers, with austenitic stainless steel, brass or bronze strike plates.	Periodic lubrication in accordance with manufacturer's instructions.
20	Mortice locks/latches complying with BS 5872 Use Category B (heavy duty). All exposed parts of austenitic stainless steel, bronze or brass and lock case (within door) of galvanized steel. Latch and deadbolt tongues of austenitic stainless steel, brass or brass with hardened steel rollers, with austenitic stainless steel, brass or bronze strike plates.	Periodic lubrication in accordance with manufacturer's instructions.
20	Rim locks complying with BS 5872 Use Category B (heavy duty). Cylinder bodies of cast brass or cast steel, with cover plate and 'pull' of brass, and galvanized or zinc plated lock case.	Periodic lubrication in accordance with manufacturer's instructions.
15	Mortice locks/latches complying with ANSI A156.2. All exposed parts of austenitic stainless steel, bronze or brass and lock case (within door) of galvanized steel. Latch and deadbolt tongues of austenitic stainless steel, brass or brass with hardened steel rollers, with austenitic stainless steel, brass or bronze strike plates.	Periodic lubrication in accordance with manufacturer's instructions.
15	Mortice locks/latches complying with BS 5872 Use Category A (domestic). All exposed parts of austenitic stainless steel, bronze or brass and lock case (within door) of galvanized steel. Latch and deadbolt tongues of austenitic stainless steel, brass or brass with hardened steel rollers, with austenitic stainless steel, brass or bronze strike plates.	Periodic lubrication in accordance with manufacturer's instructions.
15	Rim locks complying with BS 5872 Use Category A (domestic). Cylinder bodies of cast brass or cast steel, with cover plate and 'pull' of brass, and galvanized or zinc plated lock case.	Periodic lubrication in accordance with manufacturer's instructions.
10	Mortice locks/latches, with exposed parts of materials other than austenitic stainless steel, bronze or brass.	Periodic lubrication in accordance with manufacturer's instructions.

For Adjustment factors, Assumptions, Notes, Key failure modes and Key durability issues please see overleaf.

4.3d

Door ironmongery

Adjustment factors

None.

Assumptions

Locks/latches to be selected for the appropriate BS 5872 use category.

Thief resistant locks to be Kitemarked to BS 3621.

Fixings should be protectively treated (eg zinc plated) or of inherently corrosion resistant materials such as stainless steel, brass or aluminium. Fixings to be compatible with the lock/latch material, ie to prevent galvanic corrosion. For further detailed guidance, see BS PD 6484.

Notes

Locks/latches to BS 5872 and BS 3621 are marked with the manufacturer's identity and the BS number.

A number of European Standards for ironmongery are currently under development which will include a standard coding system for performance characteristics such as durability and corrosion resistance. This page will be amended to take account of the new standards once they are introduced.

Key failure modes

Failure of locking mechanism, corrosion, wear, binding.

Key durability issues

Base metal(s), surface protection, frequency of use, installation/alignment, maintenance frequency.

Door ironmongery

Description

General		Maintenance
	Mechanical push-button locks	
20	Cast brass bodies, brass knob or lever handle, with latch or deadbolts of austenitic stainless steel, brass or brass with hardened steel rollers. Bodies and operating buttons lacquered or chromium plated.	Periodic lubrication in accordance with manufacturer's instructions.
15	Bodies formed from zinc alloy diecastings, with latch or deadbolts of austenitic stainless steel, brass or brass with hardened steel rollers.	Periodic lubrication in accordance with manufacturer's instructions.

Adjustment factors

None.

Assumptions

Fixings should be protectively treated (eg zinc plated) or of inherently corrosion resistant materials such as stainless steel, brass or aluminium. Fixings to be compatible with the lock/latch material, ie to prevent galvanic corrosion. For further detailed guidance, see BS PD 6484.

Key failure modes

Corrosion, wear, failure of mechanism.

Key durability issues

Base metal, surface protection, frequency of use, degree of care in use, maintenance frequency.

4.3e

Window ironmongery

Scope

This section provides data on ironmongery for windows in non-domestic building types. It includes butt and friction hinges, espagnolette and shoot bolt mechanisms, and casement stays and fasteners.

The following component sub-types are included within this section:

	Page
• Butt hinges	4.4
• Friction hinges	4.4a
• Espagnolette/shoot bolts	4.4b
• Casement stays & wedge action fasteners	4.4c

It should be noted that there is an enormous range of different types and compositions of window ironmongery available to UK specifiers. This section aims to cover the most commonly used materials and types.

Standards cited

BS 1344:	Methods for testing vitreous enamel finishes.
Part 2:1975(1994)	Resistence to citric acid at room temperature.
BS 1615:1987(1994)	Method for specifying anodic oxidation coatings on aluminium and its alloys.
BS 6262:1994	Glazing for buildings.

BS 6375:	Performance of windows.
Part 2:1987(1995)	Specification for operation and strength characteristics.
BS 6462:1985(1994)	Specification for mechanical performance of peg-type casement stays and face-fixed wedge-action fasteners.
BS 7352:1990	Specification for strength and durability performance of metal hinges for side hanging applications and dimensional requirements for template drilled hinges.
BS 7479:1991	Method for salt spray corrosion tests in artificial atmospheres.
BS PD 6484:1979(1990)	Commentary on corrosion at bimetallic contacts and its alleviation.
BS EN 1670:1998	Building hardware - corrosion resistance - requirements and test methods.

Other references/information sources

Association of Builders' Hardware Manufacturers Codes of Practice.

Guild of Architectural Ironmongers Guidance Notes.

PSA Method of Building Product Selection Guides.

LOCATIONS - General

4 - Doors, Windows & Joinery

General	Description	Maintenance
	Butt hinges	
25	Hinges to Class 3 or better of BS 7352, with brass hinge flaps and austenitic stainless steel pins. Hinges to be painted in-situ or to comply with appropriate corrosion resistance class of BS EN 1670 (see assumptions).	Annual lubrication. If site painted, redecorate every 5 years.
20	Hinges to Class 3 or better of BS 7352, with galvanized steel hinge flaps and austenitic stainless steel pins. Hinges to be painted in-situ or to comply with appropriate corrosion resistance class of BS EN 1670 (see assumptions).	Annual lubrication. If site painted, redecorate every 5 years.
15	Hinges to Class 3 or better of BS 7352. Hinges to be painted in-situ or to comply with appropriate corrosion resistance class of BS EN 1670 (see assumptions).	Annual lubrication. If site painted, redecorate every 5 years.
10	Hinges with brass hinge flaps and mild steel pins. Hinges to be painted in-situ or to comply with appropriate corrosion resistance class of BS EN 1670 (see assumptions).	Annual lubrication. If site painted, redecorate every 5 years.
10	Hinges with mild steel hinge flaps and brass pins. Hinges to be painted in-situ or to comply with appropriate corrosion resistance class of BS EN 1670 (see assumptions).	Annual lubrication. If site painted, redecorate every 5 years.
U1	Unclassified, ie hinges with mild steel hinge flaps and mild steel pins, or hinges not in accordance with appropriate BS EN 1670 corrosion resistance class.	Unclassified.

Adjustment factors

Hinges less than class 3 corrosion resistance to BS EN 1670: -5 years.

Site painted hinges in industrial/polluted/marine environment: -5 years.

Assumptions

Unplated steel fixings should only be used where the hinge is to be painted in-situ. Otherwise, fixings should be protectively treated (eg zinc plated) or of inherently corrosion resistant materials such as stainless steel, brass or aluminium. Fixings to be compatible with the hinge material, ie to prevent galvanic corrosion. For further detailed guidance, see BS PD 6484.

Ironmongery to be tested to comply with appropriate corrosion resistance class to BS EN 1670, ie Class 1: dry internal environment; Class 2: occasionally damp interiors, areas subject to condensation; Class 3: environments that are often wet or subject to slight pollution (ie most external environments), particularly damp interiors; Class 4: very polluted environments, eg combinations of industrial and coastal pollution.

Notes

Hinges to BS 7352 must be marked with the manufacturer's identity, the BS number and hinge class.

Corrosion resistance beyond BS EN 1670 Class 4 may be required in highly polluted environments.

Key failure modes

Corrosion, wear, misalignment, distortion.

Key durability issues

Base metal, surface protection, frequency of use, degree of care in use, size/specification of hinge, maintenance frequency.

4.4

Window ironmongery

LOCATIONS - General

BPG 4 - Doors, Windows & Joinery

General	Description	Maintenance
	Friction hinges	
25	Hinges conforming to Class A of BS 6375:Part 2, with austenitic stainless steel arms, low friction slider, low friction washers at all pivot points, plastic end caps and, for side hung windows, riser blocks. Hinges to appropriate corrosion resistance class to BS EN 1670 (see assumptions).	Annual cleaning and lubrication.
25	Hinges conforming to Class A of BS 6375:Part 2, with austenitic stainless steel support arm and all other bars and tracks of extruded aluminium. Low friction slider, low friction washers at all pivot points, plastic end caps and, for side hung windows, riser blocks. Hinges to appropriate corrosion resistance class to BS EN 1670 (see assumptions).	Annual cleaning and lubrication.
20	Hinges conforming to Class A of BS 6375:Part 2, with ferritic stainless steel arms, low friction slider, low friction washers at all pivot points, plastic end caps and, for side hung windows, riser blocks. Hinges to appropriate corrosion resistance class to BS EN 1670 (see assumptions).	Annual cleaning and lubrication.
15	Hinges conforming to Class A of BS 6375:Part 2, with zinc plated and chromated steel arms, low friction slider, low friction washers at all pivot points, plastic end caps and, for side hung windows, riser blocks. Hinges to appropriate corrosion resistance class to BS EN 1670 (see assumptions).	Annual cleaning and lubrication.
U1	Unclassified, ie hinges not complying with Class A of BS 6375:Part 2, or less than above specifications.	Unclassified.

For Adjustment factors, Assumptions, Notes, Key failure modes and Key durability issues please see overleaf.

4.4a

Window ironmongery

LOCATIONS - General

Adjustment factors

Hinges not provided with low friction washers at all pivot points: life limited to 10 years.

Assumptions

Fixings to be austenitic stainless steel and appropriate for the frame material.

Ironmongery to be tested to comply with appropriate corrosion resistance class to BS EN 1670, ie Class 1: dry internal environment; Class 2: occasionally damp interiors, areas subject to condensation; Class 3: environments that are often wet or subject to slight pollution (ie most external environments), particularly damp interiors; Class 4: very polluted environments, eg combinations of industrial and coastal pollution.

Windows to be glazed in accordance with BS 6262.

It is essential that the appropriate hinge is selected for the size and weight of casement. Reference should be made to manufacturer's guidance.

The correct alignment of the casement in the opening, and of other ironmongery such as espagnolette bolts is essential to the performance of the friction hinge.

Notes

The correct installation of setting and location blocks and spacers around glazing (to prevent the window frame from dropping) is particularly important to the performance of friction hinges on side hung casements.

BS 6375:Part 2 defines two strength classes with respect to the jamming of hinges. Class A is suitable for any situation. Class B is suitable for light duty applications only, eg private domestic dwellings.

Corrosion resistance beyond BS EN 1670 Class 4 may be required in highly polluted environments.

Key failure modes

Corrosion (hinge body/fixings), wear, distortion, binding/impaired use.

Key durability issues

Type/grade of metal used in hinge/fixings, regular maintenance, correct specification (ie for size/weight of casement), fixing type.

LOCATIONS - General

General	Description	Maintenance
	Espagnolette/shoot bolts	
15	Shoot bolts (ie bolts at either end) manufactured from austenitic or ferritic stainless steel, zinc plated mild steel or plated zinc. Used with friction hinges fitted with riser plates. Bolts to comply with appropriate corrosion resistance class to BS EN 1670 (see assumptions).	Annual lubrication and adjustment (if required).
10	Espagnolette bolts (ie bolts along the side) manufactured from austenitic or ferritic stainless steel, zinc plated mild steel or plated zinc, and engaging with slotted keep plates. Used with friction hinges fitted with riser plates. Bolts to comply with appropriate corrosion resistance class to BS EN 1670 (see assumptions).	Annual lubrication and adjustment (if required).
U1	Espagnolette bolts manufactured from martenstic stainless steel, or not to appropriate BS EN 1670 corrosion resistance class.	Unclassified.

Adjustment factors

Use with friction hinges not fitted with riser plates: -5 years.

Assumptions

Installation and adjustment/alignment in strict accordance with manufacturer's instructions.

Ironmongery to be tested to comply with appropriate corrosion resistance class to BS EN 1670, ie Class 1: dry internal environment; Class 2: occasionally damp interiors, areas subject to condensation; Class 3: environments that are often wet or subject to slight pollution (ie most external environments), particularly damp interiors; Class 4: very polluted environments, eg combinations of industrial and coastal pollution.

Fixings should be protectively treated (eg zinc plated) or of inherently corrosion resistant materials such as stainless steel, brass or aluminium. Fixings to be compatible with the lock/latch material, ie to prevent galvanic corrosion. For further detailed guidance, see BS PD 6484.

Windows to be glazed in accordance with BS 6262.

It is assumed that the espagnolette/shoot bolt is used with friction hinges.

The appropriately sized hinge must be selected for the casement. Reference should be made to manufacturer's guidance.

The correct alignment of the casement in the opening, and of other ironmongery such as espagnolette bolts is essential to the performance of the friction hinge.

Notes

Espagnolette bolts require more accurate alignment with the lock keeps than do shoot bolts, and are more easily damaged if forced.

The use of thermoplastic riser plates with the friction hinges prevents side hung casements from dropping in service and causing misalignment of the lock.

Corrosion resistance beyond BS EN 1670 Class 4 may be required in highly polluted environments.

Key failure modes

Failure of locking mechanism/gears, binding, wear, misalignment, distortion.

Key durability issues

Installation, alignment, improper use (eg forcing), friction hinge specification, frequency of adjustment/maintenance.

4.4b

BPG | **4 - Doors, Windows & Joinery**

Internal	Description	Maintenance
	Casement stays & wedge action fasteners	
35	Grade 304/316 austenitic stainless steel stays/fasteners, complying with the mechanical performance requirements of BS 6462. Ironmongery to appropriate corrosion resistance class to BS EN 1670 (see assumptions).	Periodic cleaning & lubrication in accordance with manufacturer's recommendations.
30	Brass stays/fasteners, complying with the mechanical performance requirements of BS 6462. Ironmongery to appropriate corrosion resistance class to BS EN 1670 (see assumptions).	Periodic cleaning & lubrication in accordance with manufacturer's recommendations.
25	Aluminium alloy stays/fasteners, anodised to BS 1615, minimum 15 micron coating thickness, complying with the mechanical performance requirements of BS 6462. Ironmongery to appropriate corrosion resistance class to BS EN 1670 (see assumptions).	Periodic cleaning & lubrication in accordance with manufacturer's recommendations.
25	Aluminium alloy stays/fasteners, polyester powder coated to BS 6496, minimum 25 micron coating thickness, complying with the mechanical performance requirements of BS 6462. Ironmongery to appropriate corrosion resistance class to BS EN 1670 (see assumptions).	Periodic cleaning & lubrication in accordance with manufacturer's recommendations.
25	Mild steel stays/fasteners with vitreous enamel coating to Class B of BS 1344:Part 2, complying with the mechanical performance requirements of BS 6462. Ironmongery to appropriate corrosion resistance class to BS EN 1670 (see assumptions).	Periodic cleaning & lubrication in accordance with manufacturer's recommendations.
25	Diecast zinc stays/fasteners, epoxy coated, complying with the mechanical performance requirements of BS 6462. Ironmongery to appropriate corrosion resistance class to BS EN 1670 (see assumptions).	Periodic cleaning & lubrication in accordance with manufacturer's recommendations.
20	Diecast zinc stays/fasteners, nickel and chromium plated, complying with the mechanical performance requirements of BS 6462. Ironmongery to appropriate corrosion resistance class to BS EN 1670 (see assumptions).	Periodic cleaning & lubrication in accordance with manufacturer's recommendations.
15	Brass plated steel stays/fasteners, complying with the mechanical performance requirements of BS 6462. Ironmongery to appropriate corrosion resistance class to BS EN 1670 (see assumptions).	Periodic cleaning & lubrication in accordance with manufacturer's recommendations.
15	Diecast zinc stays/fasteners, nickel plated, complying with the mechanical performance requirements of BS 6462. Ironmongery to appropriate corrosion resistance class to BS EN 1670 (see assumptions).	Periodic cleaning & lubrication in accordance with manufacturer's recommendations.
10	Stays/fasteners not complying with the mechanical performance requirements of BS 6462. Ironmongery to appropriate corrosion resistance class to BS EN 1670 (see assumptions).	Periodic cleaning & lubrication in accordance with manufacturer's recommendations.

For Adjustment factors, Assumptions, Notes, Key failure modes and Key durability issues please see overleaf.

4.4c

Window ironmongery

LOCATIONS - Internal

Adjustment factors

None.

Assumptions

Fixings to be compatible with the lock/latch material, ie to prevent galvanic corrosion. For further detailed guidance, see BS PD 6484. Ironmongery to be tested to comply with appropriate corrosion resistance class to BS EN 1670, ie Class 1: dry internal environment; Class 2: occasionally damp interiors, areas subject to condensation; Class 3: environments that are often wet or subject to slight pollution (ie most external environments), particularly damp interiors; Class 4: very polluted environments, eg combinations of industrial and coastal pollution.

Notes

There is a vast range of materials and surface finishes available for window furniture. The above descriptions are indicative only.

Key failure modes

Corrosion, wear, misalignment.

Key durability issues

Base metal, surface protection, frequency of use, degree of care in use, maintenance frequency.

4.4c

Door closers

Scope

This section provides data on floor mounted, overhead and middle rail door closers for use in non-domestic building types. Door closers for use in domestic buildings are covered in the HAPM Component Life Manual and are excluded from this section.

The following component sub-types are included within this section:

	Page
• Floor mounted/overhead/middle rail	4.5

Standards cited

BS EN 1154: 1997	Building hardware-controlled door closing devices-requirements and test methods

Other references/information sources

Guild of Architectural Ironmongers.

HAPM Component Life Manual page 5.2 (domestic door closers).

Door closers

LOCATIONS - General

Description

General

Floor mounted/overhead/middle rail

20	Concealed floor-mounted door closers to BS EN 1154, incorporating 'back check' device.
15	Concealed floor-mounted door closers to BS EN 1154, not incorporating 'back check' device.
15	Overhead door closers to BS EN 1154, incorporating 'back check' device.
10	Overhead door closers to BS EN 1154, not incorporating 'back check' device.
5	Middle rail door closers.
U1	Floor-mounted/overhead door closers not to BS EN 1154.

Maintenance

Lubrication and adjustment of closing mechanism at 6 month intervals or as defined by manufacturer.

Lubrication and adjustment of closing mechanism at 6 month intervals or as defined by manufacturer.

Lubrication and adjustment of closing mechanism at 6 month intervals or as defined by manufacturer. Continuity of fixings to be checked annually.

Lubrication and adjustment of closing mechanism at 6 month intervals or as defined by manufacturer. Continuity of fixings to be checked annually.

Lubrication and adjustment of closing mechanism at 6 month intervals or as defined by manufacturer.

Unclassified.

Adjustments

Use in conjunction with 'hold open' device: +5 years (floor-mounted/overhead types only).

Closers fitted to doors expected to receive unusually frequent rough use (eg impact from trolleys/equipment): -5 years.

Assumptions

Door closer to be selected to match door leaf size and mass using Table 1 of BS EN 1154.

Door closers to be of the appropriate BS EN 1154 corrosion resistance grade for the environmental conditions (eg internal/external use).

Door closers to BS EN 1154 power sizes 3 or 4 to be specified for doors closing from 105 degrees open or 180 degrees open respectively.

Closers for fire doors must be of at least power size 3.

Door closer to be selected and specified in consultation with manufacturer.

Installation in accordance with manufacturer's instructions.

Notes

Table 1 of BS EN 1154 defines 7 power sizes of door closer, based on door leaf width and mass.

Annexe A of BS EN 1154 specifies a number of additional requirements for door closers used on fire/smoke door assemblies.

See HAPM Component Life Manual P.5.2 for domestic door closers.

Key failure modes

Stiffness, excessive/inadequate operating force, fixing failure, misalignment/misadjustment, fluid leakage.

Key durability issues

Incorporation of 'back check' device, intensity/ roughness of use, appropriate selection for use conditions (ie environment, mass of door leaf), maintenance frequency, accuracy of alignment/ adjustment.

Stairs and Balustrades

Stairs and Balustrades

BPG

Stairs

Scope

This section provides data on internal and external stairs, including fire escape stairs, for non-domestic building types. It includes separate sections on the stair structure and treads/walkways. Timber domestic stairs are covered in the HAPM Component Life Manual and are excluded from this section.

The following component sub-types are included within this section:

			Page
•	Stairs - structure:	Precast reinforced concrete	5.1
		Insitu reinforced concrete	5.1
		Aluminium	5.1a
		Steel	5.1a
•	Metal treads/walkways:	Aluminium	5.2
		Steel	5.2

Standards cited

BS 449	Specification for the use of structural steel in building
Part 2:1969	Metric units
BS 729:1971 (1994)	Specification for hot-dip galvanised coatings on iron and steel articles
BS 970	Specification for wrought steels and mechanical and allied engineering purposes
Part 1:1996	General inspection and testing procedures and specific requirements for carbon, carbon manganese, alloy and stainless steels
BS 1449	Steel plate, sheet and strip
Part 1: 1991	Carbon and carbon-manganese plate, sheet and strip
Part 2: 1983	Specification for stainless and heat resisting steel plate, sheet and strip
BS 1471: 1972	Specification for wrought aluminium and aluminium alloys for general engineering purposes - drawn tube
BS 1474 : 1987	Specification for wrought aluminium and aluminium alloy for general engineering purposes
BS 1615: 1987 (1994)	Method for specifying anodic oxidation coatings on aluminium and its alloys
BS 2994 :1976 (1987)	Specification for cold rolled steel sections
BS 4300: (various)	Wrought aluminium and aluminium alloys for general engineering purposes (supplementary series)
BS 4449: 1997	Specification for carbon steel bars for the reinforcement of concrete
BS 4592	Industrial type flooring, walkways and stair treads
Part 1: 1995	Specification for open bar gratings
Part 3: 1987	Specification for cold formed planks
Part 5: 1995	Specification for solid plates in steel, aluminium and glass reinforced plastics
BS 5395: (various)	Stairs, ladders and walkways
BS 8110	Structural use of concrete
Part 1: 1997	Code of Practice for design and construction
BS 8118:1991	Structural use of aluminium
BS EN 485: 1994	Aluminium and aluminium alloys. Sheet, strip and plate
BS EN 10025: 1993	Hot rolled products of non-alloy structural steels. Technical delivery conditions.
BS EN 10142: 1991	Specification for continuously hot-dip zinc coated low carbon steel sheet and strip for cold forming: technical delivery conditions
BS EN 10147: 1992	Specification for continuously hot-dip zinc coated structural steel sheet and strip. Technical delivery conditions
BS EN 10210:various	Hot finished structural hollow sections of non-alloy and fine grain structural steels
BS PD 6484: 1979 (1990)	Commentary on corrosion at bimetallic contacts and its alleviation

Stairs - structure _____

LOCATIONS - Internal, External

Internal	External	Description	Maintenance
		Precast reinforced concrete	
35+	35+	Precast stairs designed in accordance with BS 5395 and BS 8110:Part 1, Table 3.3 for severe exposure, with minimum cement content of 325kg/m³ with 40mm nominal cover to all reinforcement, or 400kg/m³ for 25mm reinforcement cover.	None.
35+	30	Precast stairs designed in accordance with BS 5395 and BS 8110:Part 1, Table 3.3 for moderate exposure, with minimum cement content of 300kg/m³ with 35mm nominal cover to all reinforcement, or 400kg/m³ for 20mm reinforcement cover.	None.
35	15	Precast stairs designed in accordance with BS 5395 and BS 8110:Part 1, Table 3.3 for mild exposure, with minimum cement content of 275kg/m³ with 25mm nominal cover to all reinforcement, or 300kg/m³ for 20mm reinforcement cover.	None.
U1	U1	Unclassified, ie precast stairs not complying with BS 5395 and/or BS 8110:Part 1, Table 3.3.	Unclassified.
		Insitu-reinforced concrete	
35+	30	Insitu stairs designed in accordance with BS 5395 and BS 8110:Part 1, Table 3.3 for severe exposure, with minimum cement content of 325kg/m³ with 40mm nominal cover to all reinforcement, or 400kg/m³ for 25mm reinforcement cover.	None.
35	25	Insitu stairs designed in accordance with BS 5395 and BS 8110:Part 1, Table 3.3 for moderate exposure, with minimum cement content of 300kg/m³ with 35mm nominal cover to all reinforcement, or 400kg/m³ for 20mm reinforcement cover.	None.
30	10	Insitu stairs designed in accordance with BS 5395 and BS 8110:Part 1, Table 3.3 for mild exposure, with minimum cement content of 275kg/m³ with 25mm nominal cover to all reinforcement, or 300kg/m³ for 20mm reinforcement cover.	None.
U1	U1	Unclassified, ie insitu stairs not complying with BS 5395 and/or BS 8110:Part 1, Table 3.3.	Unclassified.

For Adjustment factors, Assumptions, Notes, Key failure modes and Key durability issues please see overleaf.

5.1

Stairs - structure _____

Adjustment factors

Industrial/polluted/marine environment: -5 years.

Additional surface protection (ie anti-carbonation paint): +10 years.

Assumptions

Reinforcement steel is to BS 4449.

Cement content and reinforcement cover selected in accordance with Table 3.3 (or 4.8 for prestressed) of BS 8110:Part 1, having regard for the exposure level to be encountered in service.

Precast stairs with mechanical damage or cracking must not be installed. Precast stairs must have adequate bearing, in accordance with manufacturers' directions.

Avoid exposure to de-icing salts.

Notes

Recommendations of BS 8110 for quality of concrete should be followed.

In areas subject to salt spray components should be designed for very severe exposure to BS 8110:Part 1, Tables 3.3 or 4.8.

Domestic stairs (including timber/timber based) will be included in a future update to the HAPM Component Life Manual.

Key failure modes

Reinforcement corrosion, cracking, spalling.

Key durability issues

Reinforcement type, concrete cover, cement content, exposure conditions.

Stairs - structure

LOCATIONS - Internal, External

Internal	External	Description	Maintenance
Aluminium			
35+	35	Aluminium to BS 1474 (extrusions), BS EN 485 (fabrications, sheet), BS 1471 (drawn tube), or BS 4300 series. PVC/Plastisol coating, 200 microns nominal thickness.	Regular cleaning with non-alkaline detergent (polluted/marine areas every 3 months, other areas every 6 months). Coatings may require redecoration after 10-15 years.
35+	30	Aluminium to BS 1474 (extrusions), BS EN 485 (fabrications, sheet), BS 1471 (drawn tube), or BS 4300 series. Acrylic, PVDF/PVF2, alkyd, polyester or silicone modified polyester coating, 25-50 microns nominal thickness.	Regular cleaning with non-alkaline detergent (polluted/marine areas every 3 months, other areas every 6 months). Coatings may require redecoration after 10-15 years.
35+	30	Aluminium to BS 1474 (extrusions), BS EN 485 (fabrications, sheet), BS 1471 (drawn tube), or BS 4300 series. Anodized to BS 1615, minimum 25 micron coating.	Regular cleaning with non-alkaline detergent (polluted/marine areas every 3 months, other areas every 6 months).
35	20	Aluminium to BS 1474 (extrusions), BS EN 485 (fabrications, sheet), BS 1471 (drawn tube), or BS 4300 series. Anodized to BS 1615, minimum 10 micron coating.	Regular cleaning with non-alkaline detergent (polluted/marine areas every 3 months, other areas every 6 months).
30	15	Aluminium to BS 1474 (extrusions), BS EN 485 (fabrications, sheet), BS 1471 (drawn tube), or BS 4300 series. Mill finished.	Regular cleaning with non-alkaline detergent (polluted/marine areas every 3 months, other areas every 6 months).
U1	U1	Unclassified, ie aluminium not to relevant BS.	Unclassified.
Steel			
35+	35+	Austenitic stainless steel to BS 1449:Part 2 or BS 970:Part 1, Grade 304 or 316.	Clean every 6 months with mild detergent.
35+	30	Post galvanized mild steel, minimum 610g/m² zinc coating weight.	If painted, redecorate every 5 years.
35+	30	Pre-galvanized mild steel, minimum 610g/m² zinc coating weight. Factory applied organic coating, 25-50 microns nominal thickness.	Coatings may require redecoration after 10-15 years.
35	25	Pre-galvanized or post-galvanized mild steel, minimum 450g/m² zinc coating weight. Factory applied organic coating, 25-50 microns nominal thickness.	Coatings may require redecoration after 10-15 years.
35	25	Pre-galvanized or post-galvanized mild steel, minimum 275g/m² zinc coating weight. Factory applied PVC/Plastisol coating, 200 microns nominal thickness.	Coatings may require redecoration after 10-15 years.
30	20	Pre-galvanized or post-galvanized mild steel, minimum 275g/m² zinc coating weight. Factory applied organic coating, 25-50 microns nominal thickness.	Coatings may require redecoration after 10-15 years.
30	10	Mild steel, protected with an appropriate Micaceous Iron Oxide, Chlorinated Rubber or similar high performance finish to give a minimum dry film thickness of 250 microns.	Redecorate every 5 years.
U1	U1	Unclassified, ie mild steel 'factory primed', or less than above specifications.	Unclassified.

5.1a   **For Adjustment factors, Assumptions, Notes, Key failure modes and Key durability issues please see overleaf.**

Stairs - structure

Adjustment factors

External use in industrial/polluted /marine environment: -5 years (except 316 grade austenitic stainless steel).

Assumptions

Design in accordance with BS 5395, BS 8118 (aluminium) and BS 449 (steel).

In external/damp locations, avoid direct contact between aluminium alloys and timber treated with copper, zinc or mercury based preservatives, Oak, Sweet Chestnut, Douglas Fir, Western Red Cedar, copper alloys (or rainwater run off from), concrete, mortar or soil.

Mild steel to BS EN 10147 (post-galvanized sheet), BS EN 10142 (pre-galvanized sheet), BS 1449:Part 1, BS 970:Part 1, BS 4848 or BS 2994 (hot rolled sections).

Organic coatings include acrylic, PVDF/PVF2, alkyd, polyester and silicone modified polyester (steel only).

Hot dip galvanizing to BS 729 (steel only).

With galvanized hollow sections it is essential to provide holes for venting and drainage, and to ensure that the internal surfaces are fully coated (steel only).

The lives quoted represent the life to first maintenance. Whether maintenance is possible or even probable will depend on the feasibility of access to ALL surfaces. In most cases such access will not be possible.

Where two or more metals are used in the stair construction, they should be compatible, ie to prevent galvanic corrosion. For further detailed guidance, see BS PD 6484.

Notes

Domestic stairs (including timber/timber based) will be included in a future update to the HAPM Component Life Manual.

Key failure modes

Corrosion, scratching/chalking/peeling of surface finish, impact/indentation/distortion.

Key durability issues

Type/thickness of surface protection, thickness/profile of metal, maintenance frequency, exposure conditions.

Stairs - metal treads/walkways

5 - Stairs & Balustrades

LOCATIONS - General

Internal	External	Description	Maintenance
Aluminium			
35	30	Aluminium to BS 1474 (extrusions), BS EN 485 (fabrications, sheet), BS 1471 (drawn tube), or BS 4300 series. PVC/Plastisol coating, 200 microns nominal thickness.	Regular cleaning with non-alkaline detergent (polluted/marine areas every 3 months, other areas every 6 months). Coatings may require redecoration after 5-10 years depending on use.
35	25	Aluminium to BS 1474 (extrusions), BS EN 485 (fabrications, sheet), BS 1471 (drawn tube), or BS 4300 series. Acrylic, PVDF/PVF2, alkyd, polyester or silicone modified polyester coating, 25-50 microns nominal thickness.	Regular cleaning with non-alkaline detergent (polluted/marine areas every 3 months, other areas every 6 months). Coatings may require redecoration after 5-10 years depending on use.
35	25	Aluminium to BS 1474 (extrusions), BS EN 485 (fabrications, sheet), BS 1471 (drawn tube), or BS 4300 series. Anodized to BS 1615, minimum 25 micron coating.	Regular cleaning with non-alkaline detergent (polluted/marine areas every 3 months, other areas every 6 months).
30	15	Aluminium to BS 1474 (extrusions), BS EN 485 (fabrications, sheet), BS 1471 (drawn tube), or BS 4300 series. Anodized to BS 1615, minimum 10 micron coating.	Regular cleaning with non-alkaline detergent (polluted/marine areas every 3 months, other areas every 6 months).
25	10	Aluminium to BS 1474 (extrusions), BS EN 485 (fabrications, sheet), BS 1471 (drawn tube), or BS 4300 series. Mill finished.	Regular cleaning with non-alkaline detergent (polluted/marine areas every 3 months, other areas every 6 months).
U1	U1	Unclassified, ie aluminium not to relevant BS.	Unclassified.
Steel			
35	30	Austenitic stainless steel to BS 1449:Part 2 or BS 970:Part 1, Grade 304 or 316.	Clean every 6 months with mild detergent.
35	25	Post galvanized mild steel, minimum 610g/m² zinc coating weight.	If painted, redecorate every 3-5 years depending on use.
35	25	Pre-galvanized mild steel, minimum 610g/m² zinc coating weight. Factory applied organic coating, 25-50 microns nominal thickness.	Coatings may require redecoration after 5-10 years depending on use.
30	20	Pre-galvanized or post-galvanized mild steel, minimum 450g/m² zinc coating weight. Factory applied organic coating, 25-50 microns nominal thickness.	Coatings may require redecoration after 5-10 years depending on use.
30	20	Pre-galvanized or post-galvanized mild steel, minimum 275g/m² zinc coating weight. Factory applied PVC/Plastisol coating, 200 microns nominal thickness.	Coatings may require redecoration after 5-10 years depending on use.
25	15	Pre-galvanized or post-galvanized mild steel, minimum 275g/m² zinc coating weight. Factory applied organic coating, 25-50 microns nominal thickness.	Coatings may require redecoration after 5-10 years depending on use.
25	5	Mild steel, protected with an appropriate Micaceous Iron Oxide, Chlorinated Rubber or similar high performance finish to give a minimum dry film thickness of 250 microns.	Redecorate every 3-5 years depending on use.
U1	U1	Unclassified, ie mild steel 'factory primed', or less than above specifications.	Unclassified.

5.2   **For Adjustment factors, Assumptions, Notes, Key failure modes and Key durability issues please see overleaf.**

Stairs - metal treads/walkways _

LOCATIONS - General

Adjustment factors

External use in industrial/polluted/marine environment: -5 years (except 316 grade austenitic stainless steel).

Stairs in occasional use only (eg fire escape stairs): +5 years.

Assumptions

Design in accordance with BS 4592:Part 1 (open bar gratings), Part 3 (cold formed planks), Part 5 (solid plates).

Design in accordance with appropriate BS 4592 loading category (light, general, heavy duty).

In external/damp locations, avoid direct contact between aluminium alloys and timber treated with copper, zinc or mercury based preservatives, Oak, Sweet Chestnut, Douglas Fir, Western Red Cedar, copper alloys (or rainwater run off from), concrete, mortar or soil.

Mild steel to BS EN 10147 (post-galvanized sheet), BS EN 10142 (pre-galvanized sheet), BS EN 10025 (low carbon steel for open bar gratings, solid plates).

Organic coatings include acrylic, PVDF/PVF2, alkyd, polyester and silicone modified polyester.

Hot dip galvanizing to BS 729. Open bar gratings to BS 4592:Part 1 must be free draining (steel only).

The lives quoted represent the life to first maintenance. Whether maintenance is possible or even probable will depend on the feasibility of access to ALL surfaces. In most cases such access will not be possible.

Where two or more metals are used in the stair construction, they should be compatible, ie to prevent galvanic corrosion. For further detailed guidance, see BS PD 6484.

Notes

Domestic stairs (including timber/timber based) will be included in a future update to the HAPM Component Life Manual.

Key failure modes

Corrosion, scratching/chalking/peeling of surface finish, impact/indentation/distortion, wear.

Key durability issues

Type/thickness of surface protection, thickness/profile of metal, maintenance frequency, intensity of use, exposure conditions.

Balustrades

Scope

This section provides data on balustrades for internal and external use in non-domestic building types. It includes metal and timber frames and metal, timber and glazed infill panels. Domestic balustrades are to be covered in the HAPM Component Life Manual and are excluded from this section.

The following component sub-types are included within this section:

		Page
•	Frames:	
	Stainless steel	5.3
	Mild steel	5.3
	Aluminium	5.3a
	Softwood (external use)	5.3b
	Hardwood (external use)	5.3c
	Softwood (internal use)	5.3d
	Hardwood (internal use)	5.3d
•	Infill panels:	
	Stainless steel	5.4
	Mild steel	5.4
	Aluminium	5.4a
	Glass	5.4b
	Plastic	5.4b
	Softwood (external use)	5.4c
	Hardwood (external use)	5.4d
	Softwood (internal use)	5.4e
	Hardwood (internal use)	5.4e
	Plywood	5.4f

Standards cited

BS 729: 1971 (1994)	Specification for hot-dip galvanised coatings on iron and steel articles
BS 970	Specification for wrought steels and mechanical and allied engineering purposes
BS 1088: 1966 (1988)	Specifications for plywood for marine craft
BS 1449	Steel plate, sheet and strip
Part 1: 1991	Carbon and carbon-manganese plate, sheet and strip
Part 2: 1983	Specification for stainless and heat resisting steel plate, sheet and strip
BS 1471: 1972	Specification for wrought aluminium and aluminium alloys for general engineering purposes - drawn tube
BS 1474: 1987	Specification for wrought aluminium and aluminium alloy for general engineering purposes
BS 1615: 1987 (1994)	Method for specifying anodic oxidation coatings on aluminium and its alloys
BS 2994: 1976 (1987)	Specification for cold rolled steel sections
BS 3416: 1991	Specification for bitumen-based coatings for cold application, suitable for use in contact with potable water
BS 3605: (various)	Austenitic stainless steel pipes & tubes for pressure purposes
BS 4203	Extruded rigid PVC corrugated sheeting
Part 1: 1980 (1994)	Specification for performance requirements
BS 4254: 1983 (1991)	Specification for two-part polysulphide-based sealants
BS 4300: (various)	Wrought aluminium and aluminium alloys for general engineering purposes
BS 6180: 1995	Code of Practice for barriers in and about buildings
BS 6206: 1981 (1994)	Specification for impact performance requirements for flat safety glass and safety plastics for use in buildings
BS 6262: 1994	Glazing for buildings
BS 6566	Plywood
Part 8: 1985 (1991)	Specification for bond performance of veneer plywood

Continued overleaf

Balustrades

Standards cited *continued*

BS EN 204: 1991
Classification of non-structural adhesives for joining of wood and derived timber products

BS EN 485: 1994
EN 636: 1996
Aluminium and aluminium alloys. Sheet, strip and plateBS Plywood. Specifications.

BS EN 942: 1996
Timber in joinery. General classification of timber quality.

BS EN 10142: 1991
Specification for continuously hot-dip zinc coated low carbon steel sheet and strip for cold forming: technical delivery conditions

BS EN 10147: 1992
Specification for continuously hot-dip zinc coated structural steel sheet and strip. Technical delivery conditions

BS EN 10210:various
Hot finished structural hollow sections of non-alloy and fine grain structural steels

PD 6484: 1979 (1990)
Commentary on corrosion at bimetallic contacts and its alleviation

Other references/information sources

BRE Digest 296:
Timbers: their natural durability and resistance to preservative treatment.

BRE Digest 323:
Selecting wood-based panel products.

British Wood Preserving Association (BWPA) Manual.

HAPM Component Life Manual pages 4.11 - 4.13 (sealants; sealed glazing units), pages 7.7 - 7.7c (domestic handrails/guardings)

TYPE

Balustrades - frames _____

LOCATIONS - Internal, External

SUB TYPES
Stainless steel
Mild steel

BPG | 5 - Stairs & Balustrades

Internal	External	Description	Maintenance
		Stainless steel	
35+	35+	Austenitic stainless steel to BS 970:Part 1/BS 3605, grade 315S16, 316S16/31/33.	Regular cleaning with mild detergent.
35+	35	Austenitic stainless steel to BS 970:Part 1/BS 3605, grade 304S16.	Regular cleaning with mild detergent.
35+	25	Ferritic stainless steel to BS 970:Part 1/BS 3605, grade 430S17, 434S17.	Regular cleaning with mild detergent.
35+	20	Stainless steel to BS 970:Part 1/BS 3605, unknown grade.	Regular cleaning with mild detergent.
U1	U1	Unclassified, ie stainless steel, not to BS 970/BS 6180, and/or not designed to BS 6180 criteria.	Unclassified.
		Mild steel	
35+	30	Post galvanized mild steel, minimum 610g/m² zinc coating weight.	If painted, redecorate every 5 years (external only).
35+	30	Pre-galvanized mild steel, minimum 610g/m² zinc coating weight. Factory applied organic coating, 25-50 microns nominal thickness.	Coatings may require redecoration after 10-15 years.
35	25	Pre-galvanized or post-galvanized mild steel, minimum 450g/m² zinc coating weight. Factory applied organic coating, 25-50 microns nominal thickness.	Coatings may require redecoration after 10-15 years.
35	25	Pre-galvanized or post-galvanized mild steel, minimum 275g/m² zinc coating weight. Factory applied PVC/Plastisol coating, 200 microns nominal thickness.	Coatings may require redecoration after 10-15 years.
30	20	Pre-galvanized or post-galvanized mild steel, minimum 275g/m² zinc coating weight. Factory applied organic coating, 25-50 microns nominal thickness.	Coatings may require redecoration after 10-15 years.
30	10	Mild steel, protected with an appropriate Micaceous Iron Oxide, Chlorinated Rubber or similar high performance finish to give a minimum dry film thickness of 250 microns.	Redecorate every 5 years.
U1	U1	Unclassified, ie mild steel 'factory primed', or less than above specifications.	Unclassified.

For Adjustment factors, Assumptions, Key failure modes and Key durability issues please see overleaf.

5.3

BPG

Balustrades - frames

Adjustment factors

Industrial/polluted/marine environment: -5 years (except 316 grade austenitic stainless steel).

Assumptions

Design and installation (including fixings) in accordance with BS 6180.

Balustrade (including frame and infill) to be designed for the appropriate use category as defined in Table 1 of BS 6180 (eg private residential, institutional, office).

Mild steel to BS EN 10147 (post-galvanized sheet), BS EN 10142 (pre-galvanized sheet), BS 1449:Part 1, BS 970:Part 1, BS 4848 or BS 2994 (hot rolled sections).

Organic coatings include acrylic, PVDF/PVF2, alkyd, polyester and silicone modified polyester.

Hot dip galvanizing to BS 729. With galvanized hollow sections for external use, it is essential to provide holes for venting and drainage, and to ensure that the internal surfaces are fully coated.

Where coated metal sections are set into concrete or into the ground, the embedded sections should be painted with a bituminous solution complying with type 1 of BS 3416.

The lives quoted represent the life to first maintenance. Whether maintenance is possible or even probable will depend on the feasibility of access to ALL surfaces. In most cases such access will not be possible.

Where two or more metals are used in the balustrade/stair construction, they should be compatible, ie to prevent galvanic corrosion. For further detailed guidance, see BS PD 6484.

Key failure modes

Corrosion, scratching/chalking/peeling of surface finish, impact/indentation/distortion.

Key durability issues

Type/thickness of surface protection, thickness/profile of metal, maintenance frequency, exposure conditions.

Balustrades - frames

LOCATIONS - Internal, External

Description

Aluminium

Internal	External	Description
35+	35	Aluminium to BS 1474 (extrusions)/BS EN 485 (fabrications/sheet), BS 1471 (drawn tube), or BS 4300 series. PVC/Plastisol coating, 200 microns nominal thickness.
35+	30	Aluminium to BS 1474 (extrusions)/BS EN 485 (fabrications/sheet), BS 1471 (drawn tube), or BS 4300 series. Acrylic, PVDF/PVF2, alkyd, polyester or silicone modified polyester coating, 25-50 microns nominal thickness.
35+	30	Aluminium to BS 1474 (extrusions)/BS EN 485 (fabrications/sheet), BS 1471 (drawn tube), or BS 4300 series. Anodized to BS 1615, minimum 25 micron coating.
35	20	Aluminium to BS 1474 (extrusions)/BS EN 485 (fabrications/sheet), BS 1471 (drawn tube), or BS 4300 series. Anodized to BS 1615, minimum 10 micron coating.
30	15	Aluminium to BS 1474 (extrusions)/BS EN 485 (fabrications/sheet), BS 1471 (drawn tube), or BS 4300 series. Mill finished.
U1	U1	Unclassified, ie aluminium, not to relevant BS.

Maintenance

Regular cleaning with non-alkaline detergent (polluted/marine areas every 3 months, other areas every 6 months). Coatings may require redecoration after 10-15 years.

Regular cleaning with non-alkaline detergent (polluted/marine areas every 3 months, other areas every 6 months). Coatings may require redecoration after 10-15 years.

Regular cleaning with non-alkaline detergent (polluted/marine areas every 3 months, other areas every 6 months).

Regular cleaning with non-alkaline detergent (polluted/marine areas every 3 months, other areas every 6 months).

Regular cleaning with non-alkaline detergent (polluted/marine areas every 3 months, other areas every 6 months).

Unclassified.

Key failure modes

Corrosion, scratching/chalking/peeling of surface finish, impact/indentation/distortion.

Key durability issues

Type/thickness of surface protection, thickness/profile of metal, maintenance frequency, exposure conditions.

Adjustment factors

Industrial/polluted/marine environment: -5 years.

Assumptions

Design and installation (including fixings) in accordance with BS 6180.

Balustrade (including frame and infill) to be designed for the appropriate use category as defined in Table 1 of BS 6180 (eg private residential, institutional, office).

In external/damp locations, avoid direct contact between aluminium alloys and timber treated with copper, zinc or mercury based preservatives, Oak, Sweet Chestnut, Douglas Fir, Western Red Cedar, copper alloys (or rainwater run off from), concrete, mortar or soil.

Where coated metal sections are set into concrete or into the ground, the embedded sections should be painted with a bituminous solution complying with type 1 of BS 3416.

The lives quoted represent the life to first maintenance. Whether maintenance is possible or even probable will depend on the feasibility of access to ALL surfaces. In most cases such access will not be possible.

Where two or more metals are used in the balustrade/ stair construction, they should be compatible, ie to prevent galvanic corrosion. For further detailed guidance, see BS PD 6484.

5.3a
Aluminium

Balustrades - frames

LOCATIONS - External

External	Description	Maintenance
	Softwood (external use)	
25	Permeable softwoods (eg Scots Pine [Redwood]) pressure impregnated with CCA to BWPA schedule P2 or with creosote to schedule T4. Non permeable softwoods (eg Douglas Fir, Hemlock, Spruce) pressure impregnated with CCA to BWPA schedule P4 or with creosote to schedule T5.	Redecorate: stain every 3 years or paint every 5 years. Renew any creosote coating every 3 years.
25	Permeable softwoods double vacuum impregnated with organic solvent to BWPA schedule V1 or V2. Non permeable softwoods, or mixed species, double vacuum impregnated with organic solvent to BWPA schedule V3.	Redecorate: stain every 3 years or paint every 5 years. Renew any creosote coating every 3 years.
20	Heartwood only of untreated softwood of a species designated as 'SW' or 'SWC' (suitable) for external use without preservation in Table NA1 of BS EN 942 (eg Western Red Cedar and American Douglas Fir).	Redecorate: stain every 3 years or paint every 5 years. Renew any creosote coating every 3 years.
10	Non permeable softwoods, or mixed species, double vacuum impregnated with organic solvent to BWPA schedule V1 or V2. Or pressure impregnated with CCA or BWPA schedule P2 or creosote to BWPA schedule T4.	Redecorate: stain every 3 years or paint every 5 years. Renew any creosote coating every 3 years.
10	Permeable softwoods dipped/immersed in organic solvent for a minimum of 3 minutes.	Redecorate: stain every 3 years or paint every 5 years. Renew any creosote coating every 3 years.
5	Non permeable softwoods, or mixed species, dipped/immersed in organic solvent for a minimum of 3 minutes.	Redecorate: stain every 3 years or paint every 5 years. Renew any creosote coating every 3 years.
5	Untreated softwoods of a species designated as 'SP' or 'SPC' (only suitable for external use if preservative treated) in Table NA1 of BS EN 942 (eg Hemlock, Scots Pine, European Redwood) or softwood only treated by brushing or untimed immersion.	Redecorate: stain every 3 years or paint every 5 years. Renew any creosote coating every 3 years.
U1	Unclassified, ie untreated softwoods of indeterminate species or of a species designated as 'X' (unsuitable) for external use in Table NA1 of BS EN 942 (eg Parana Pine).	Unclassified.

For Adjustment factors, Assumptions, Key failure modes and Key durability issues please see overleaf.

5.3b

Balustrades - frames _____

Key failure modes

Fungal/insect attack, cracks/splits, open joints, distortion, movement.

Key durability issues

Timber species & permeability, preservative treatment, surface protection, maintenance frequency, adhesive type, exposure conditions.

Designed and constructed in accordance with BS 6180.

General timber quality must be defined in the job specification either by reference to BS EN 942 (Classes J2-J50) or by a suitable structural timber specification (to limit knots, fast grown timber, grain slope and the like).

In the majority of cases specifications for structural timbers will only make reference to strength grades (eg spruce-pine-fir); GS, MGS, SC3 and the like) which will inevitably mean mixed species which will include permeable and non permeable types.

Fixings and connectors to be plated or non-ferrous.

Adhesives to be minimum type 'D3' to BS EN 204.

Adjustment factors

Stained: -5 years.

Type 'D3' adhesive to BS EN 204: -5 years.

Assumptions

All timber specifications will be deemed to mean both heartwood and sapwood unless heartwood explicitly required.

PERMEABLE species means 'easy to treat' and 'moderately easy to treat', NON PERMEABLE species means 'difficult to treat and 'extremely difficult to treat' as defined in BS EN 350-2.

Organic solvent preservatives to be to BWPA Table 4 Type F/N (ie fungicidal and Insecticidal).

References to BWPA treatment schedules means schedules included in the British Wood Preserving Association Manual.

Balustrades - frames

LOCATIONS - External

External	Description	Maintenance
	Hardwood (external use)	
30	Permeable hardwoods (eg Opepe, English Elm) pressure impregnated with CCA to BWPA schedule P2 or with creosote to schedule T4. Non permeable hardwoods (eg Idigbo. Luan, Mahogany, Meranti, European Oak, Utile) pressure impregnated with CCA to BWPA schedule P4 or with creosote to schedule T5.	Redecorate: stain every 3 years or paint every 5 years. Renew any creosote coating every 3 years.
30	Permeable hardwoods double vacuum impregnated with organic solvent to BWPA schedule V1 or V2. Non permeable hardwoods, or mixed species, double vacuum impregnated with organic solvent to BWPA schedule V3.	Redecorate: stain every 3 years or paint every 5 years. Renew any creosote coating every 3 years.
25	Heartwood only of untreated hardwood of a species designated as 'SW' or 'SWC' (suitable) for external use without preservation in Table NA2 of BS EN 942 (eg Afromosia, Iroko, Mahogany [but not 'Philippine' mahogany], European Oak, Sapele, Teak, Utile).	Redecorate: stain every 3 years or paint every 5 years. Renew any creosote coating every 3 years.
15	Non permeable hardwoods, or mixed species, double vacuum impregnated with organic solvent to BWPA schedule V1 or V2. Or pressure impregnated with CCA or BWPA schedule P2 or creosote to BWPA schedule T4.	Redecorate: stain every 3 years or paint every 5 years. Renew any creosote coating every 3 years.
15	Permeable hardwoods dipped/immersed in organic solvent for a minimum of 3 minutes.	Redecorate: stain every 3 years or paint every 5 years. Renew any creosote coating every 3 years.
10	Non permeable hardwoods, or mixed species, dipped/immersed in organic solvent for a minimum of 3 minutes.	Redecorate: stain every 3 years or paint every 5 years. Renew any creosote coating every 3 years.
5	Untreated hardwoods of a species designated as 'SP' or 'SPC' (only suitable for external use if preservative treated) in Table NA2 of BS EN 942 (eg Luan, Meranti, Agba) or hardwood only treated by brushing or untimed immersion.	Redecorate: stain every 3 years or paint every 5 years. Renew any creosote coating every 3 years.
U1	Unclassified, ie untreated hardwoods of indeterminate species or of a species designated as 'X' (unsuitable) for external use in Table NA2 of BS EN 942 (eg Ash, Beech, Ramin, Sycamore).	Unclassified.

For Adjustment factors, Assumptions, Key failure modes and Key durability issues please see overleaf.

5.3c

Balustrades - frames

Key failure modes

Fungal/insect attack, cracks/splits, open joints, distortion, movement.

Key durability issues

Timber species & permeability, preservative treatment, surface protection, maintenance frequency, adhesive type, exposure conditions.

Designed and constructed in accordance with BS 6180.

General timber quality must be defined in the job specification either by reference to BS EN 942 (Classes J2-J50) or by a suitable structural timber specification (to limit knots, fast grown timber, grain slope and the like).

In the majority of cases specifications for structural timbers will only make reference to strength grades (eg spruce-pine-fir); GS, MGS, SC3 and the like) which will inevitably mean mixed species which will include permeable and non permeable types.

Fixings and connectors to be plated or non-ferrous.

Adhesives to be minimum type 'D3' to BS EN 204.

Adjustment factors

Stained: -5 years.

Type 'D3' adhesive to BS EN 204: -5 years.

Assumptions

All timber specifications will be deemed to mean both heartwood and sapwood unless heartwood explicitly required.

PERMEABLE species means 'easy to treat', NON PERMEABLE species means 'difficult to treat' and 'extremely difficult to treat' as defined in BS EN 350-2.

Organic solvent preservatives to be to BWPA Table 4 Type F/N (ie fungicidal and insecticidal).

References to BWPA treatment schedules means schedules included in the British Wood Preserving Association Manual.

Balustrades - frames

LOCATIONS - Internal

BPG	5 - Stairs & Balustrades

Description

Internal		Maintenance
Softwood (internal use)		
35+	Softwood to BS EN 942 Table NA1 designated as suitable for use in internal joinery work.	Redecorate: stain every 5 years, paint every 10 years.
5	Unspecified timber species or quality. Not to BS EN 942 Table NA 1 and/or not suitable for internal joinery.	Redecorate: stain every 5 years, paint every 10 years.
Hardwood (internal use)		
35+	Hardwood to BS EN 942 Table NA2 designated as suitable for use in internal joinery work.	Redecorate: stain every 5 years, paint every 10 years.
5	Unspecified timber species or quality. Not to BS EN 942 Table NA2 and/or not suitable for internal joinery.	Redecorate: stain every 5 years, paint every 10 years.

Key failure modes

Fungal/insect attack, cracks/splits, open joints, distortion, movement.

Key durability issues

Timber species & permeability, preservative treatment, surface protection, maintenance frequency, adhesive type.

Adjustment factors

None.

Assumptions

Designed and constructed in accordance with BS 6180.

Joinery to be manufactured from timber dried down to the appropriate moisture content for in-service conditions, ie 13-17% for intermittent heating or 8-12% for continuous heating.

Joinery to be transported and stored under cover and well ventilated. Internal joinery must not be installed until the building is watertight.

5.3d

Balustrades - infill panels

LOCATIONS - Internal, External

BPG **5 - Stairs & Balustrades**

Internal	External	Description	Maintenance
		Stainless steel	
35+	35+	Austenitic stainless steel to BS 1449:Part 2 or BS 970:Part 1, grade 315S16, 316S16/31/33.	Regular cleaning with mild detergent.
35+	35	Austenitic stainless steel to BS 1449:Part 2 or BS 970:Part 1, grade 304S16.	Regular cleaning with mild detergent.
35+	30	Ferritic stainless steel to BS 1449:Part 2 or BS 970:Part 1, grade 430S17, 434S17.	Regular cleaning with mild detergent.
35	25	Stainless steel to BS 1449:Part 2 or BS 970:Part 1, unknown grade.	Regular cleaning with mild detergent.
U1	U1	Unclassified, ie stainless steel, not to BS 1449:Part 2 or BS 970:Part 1, and/or not designed to BS 6180 criteria.	Unclassified.
		Mild steel	
35+	30	Post galvanized mild steel, minimum 610g/m² zinc coating weight.	If painted, redecorate every 5 years (external only).
35+	30	Pre-galvanized mild steel, minimum 610g/m² zinc coating weight. Factory applied organic coating, 25-50 microns nominal thickness.	Coatings may require redecoration after 10-15 years.
35	25	Pre-galvanized or post-galvanized mild steel, minimum 450g/m² zinc coating weight. Factory applied organic coating, 25-50 microns nominal thickness.	Coatings may require redecoration after 10-15 years.
35	25	Pre-galvanized or post-galvanized mild steel, minimum 275g/m² zinc coating weight. Factory applied PVC/Plastisol coating, 200 microns nominal thickness.	Coatings may require redecoration after 10-15 years.
30	20	Pre-galvanized or post-galvanized mild steel, minimum 275g/m² zinc coating weight. Factory applied organic coating, 25-50 microns nominal thickness.	Coatings may require redecoration after 10-15 years.
30	10	Mild steel, protected with an appropriate Micaceous Iron Oxide, Chlorinated Rubber or similar high performance finish to give a minimum dry film thickness of 250 microns.	Redecorate every 5 years.
U1	U1	Unclassified, ie mild steel 'factory primed', or less than above specifications.	Unclassified.

For Adjustment factors, Assumptions, Key failure modes and Key durability issues please see overleaf.

5.4

Balustrades - framework ___

LOCATIONS - Internal, External

Adjustment factors

Industrial/polluted/marine environment: -5 years (except 316 grade austenitic stainless steel).

Assumptions

Design and installation (including fixings) in accordance with BS 6180.

Balustrade (including frame and infill) to be designed for the appropriate use category as defined in Table 1 of BS 6180 (eg private residential, institutional, office).

Mild steel to BS EN 10147 (post-galvanized sheet), BS EN 10142 (pre-galvanized sheet).

Organic coatings include acrylic, PVDF/PVF2, alkyd, polyester and silicone modified polyester.

Hot dip galvanizing to BS 729.

The lives quoted represent the life to first maintenance. Whether maintenance is possible or even probable will depend on the feasibility of access to ALL surfaces. In most cases such access will not be possible.

Where two or more metals are used in the balustrade/stair construction, they should be compatible, ie to prevent galvanic corrosion. For further detailed guidance, see BS PD 6484.

Fixings to be minimum galvanized steel or non-ferrous.

Key failure modes

Corrosion, scratching/chalking/peeling of surface finish, impact/indentation/distortion.

Key durability issues

Type/thickness of surface protection, thickness/profile of metal skin, support to skin (eg rigid core), fixing of panels in frame, maintenance frequency, exposure conditions.

Balustrades - framework

BPG | 5 - Stairs & Balustrades

Internal	External	Description
		Aluminium
35+	35	Aluminium to BS 1474 (extrusions)/BS EN 485 (fabrications/sheet). PVC/Plastisol coating, 200 microns nominal thickness.
35+	30	Aluminium to BS 1474 (extrusions)/BS EN 485 (fabrications/sheet). Acrylic, PVDF/PVF2, alkyd, polyester or silicone modified polyester coating, 25-50 microns nominal thickness.
35+	30	Aluminium to BS 1474 (extrusions)/BS EN 485 (fabrications/sheet). Anodized to BS 1615, minimum 25 micron coating.
35	20	Aluminium to BS 1474 (extrusions)/BS EN 485 (fabrications/sheet). Anodized to BS 1615, minimum 10 micron coating.
30	15	Aluminium to BS 1474 (extrusions)/BS EN 485 (fabrications/sheet). Mill finished.
U1	U1	Unclassified, ie aluminium, not to relevant BS.

Maintenance

Regular cleaning with non-alkaline detergent (polluted/marine areas every 3 months, other areas every 6 months). Coatings may require redecoration after 10-15 years.

Regular cleaning with non-alkaline detergent (polluted/marine areas every 3 months, other areas every 6 months). Coatings may require redecoration after 10-15 years.

Regular cleaning with non-alkaline detergent (polluted/marine areas every 3 months, other areas every 6 months).

Regular cleaning with non-alkaline detergent (polluted/marine areas every 3 months, other areas every 6 months).

Regular cleaning with non-alkaline detergent (polluted/marine areas every 3 months, other areas every 6 months).

Unclassified.

Key failure modes

Corrosion, scratching/chalking/peeling of surface finish, impact/indentation/distortion.

Key durability issues

Type/thickness of surface protection, thickness/profile of metal skin, support to skin (eg rigid core), fixing of panels in frame, maintenance frequency, exposure conditions.

Adjustment factors

Industrial/polluted/marine environment: -5 years.

Assumptions

Design and installation (including fixings) in accordance with BS 6180.

Balustrade (including frame and infill) to be designed for the appropriate use category as defined in Table 1 of BS 6180 (eg private residential, institutional, office).

In external/damp locations, avoid direct contact between aluminium alloys and timber treated with copper, zinc or mercury based preservatives, Oak, Sweet Chestnut, Douglas Fir, Western Red Cedar, copper alloys (or rainwater run off from), concrete, mortar or soil.

The lives quoted represent the life to first maintenance. Whether maintenance is possible or even probable will depend on the feasibility of access to ALL surfaces. In most cases such access will not be possible.

Where two or more metals are used in the balustrade/stair construction, they should be compatible, ie to prevent galvanic corrosion. For further detailed guidance, see BS PD 6484.

Fixings to be minimum galvanized steel or non-ferrous.

5.4a

Balustrades - infill panels

SUB TYPES
Glass
Plastic

5 - Stairs & Balustrades

Internal	External	Description	Maintenance
Glass			
35+	35+	Laminated glass to BS 6206. Thickness and framing/fixings in accordance with BS 6180.	Periodic cleaning with warm water and mild (non-abrasive) detergent. Inspect fixings (and tighten if necessary) annually. Inspect gaskets annually, replace as necessary.
35+	35+	Toughened glass to BS 6206. Thickness and framing/fixings in accordance with BS 6180.	Periodic cleaning with warm water and mild (non-abrasive) detergent. Inspect fixings (and tighten if necessary) annually. Inspect gaskets annually, replace as necessary.
35	35	Wired glass to BS 6206. Thickness and framing/fixings in accordance with BS 6180.	Periodic cleaning with warm water and mild (non-abrasive) detergent. Inspect fixings (and tighten if necessary) annually. Inspect gaskets annually, replace as necessary.
U1	U1	Unclassified, ie glass not to BS 6206 and/or BS 6180.	Unclassified.
Plastic			
35	25	UV stabilised GRP manufactured to BS 4254 and complying with impact performance requirements of BS 6206 (class A or C). Surfaces protected with factory applied fluoride coat or UV protected film.	Periodic cleaning with warm water and mild (non-abrasive) detergent. Application of acrylic or polyester lacquer to restore sheet surface (typically between years 5 and 10 depending upon rate of surface erosion). Inspect fixings (and tighten if necessary) annually. Inspect gaskets annually, replace as necessary.
30	20	Polycarbonate complying with impact performance requirements of BS 6206 (class A or C). Surfaces protected with factory applied UV protective film/coating.	Periodic cleaning with warm water and mild (non-abrasive) detergent. Application of acrylic or polyester lacquer to restore sheet surface (typically between years 5 and 10 depending upon rate of surface erosion). Inspect fixings (and tighten if necessary) annually. Inspect gaskets annually, replace as necessary.
30	20	Cast/extruded acrylic complying with impact performance requirements of BS 6206 (class A or C).	Periodic cleaning with warm water and mild (non-abrasive) detergent. Application of acrylic or polyester lacquer to restore sheet surface (typically between years 5 and 10 depending upon rate of surface erosion). Inspect fixings (and tighten if necessary) annually. Inspect gaskets annually, replace as necessary.
25	10	UV stabilised GRP manufactured to BS 4254 and complying with impact performance requirements of BS 6206 (class A or C). Surface protected with polyester film or gel coat.	Periodic cleaning with warm water and mild (non-abrasive) detergent. Application of acrylic or polyester lacquer to restore sheet surface (typically between years 5 and 10 depending upon rate of surface erosion). Inspect fixings (and tighten if necessary) annually. Inspect gaskets annually, replace as necessary.
20	10	UV stabilised, extruded/thermoformed PVC-u to BS 4203:Part 1 (extruded only) and complying with impact performance requirements of BS 6206 (class A or C).	Periodic cleaning with warm water and mild (non-abrasive) detergent. Application of acrylic or polyester lacquer to restore sheet surface (typically between years 5 and 10 depending upon rate of surface erosion). Inspect fixings (and tighten if necessary) annually. Inspect gaskets annually, replace as necessary.
U1	U1	Unclassified, ie plastic not to relevant BS and/or not complying with impact performance requirements of BS 6206.	Unclassified.

For Adjustment factors, Assumptions, Notes, Key failure modes and Key durability issues please see overleaf.

5.4b

Balustrades - framework _____

Adjustment factors

Use in industrial/polluted/marine environment: -5 years (plastics only).

Use in contact with Plastisol coated steel: -10 years (polycarbonate only).

Factory applied surface protection: +5 years (PVC-u only).

Assumptions

Design and installation in accordance with BS 6180 or, for low level glazing forming part of building facade, to BS 6262.

Balustrade (including frame and infill) to be designed for the appropriate use category as defined in Table 1 of BS 6180 (eg private residential, institutional, office).

Notes

Plastic materials may need to be reinforced with wire or metal mesh in order to meet the impact performance requirements of BS 6206.

Some UV discolouration/yellowing can be expected to occur in most plastics exposed to sunlight within 5-10 years.

Polycarbonate is prone to damage by solvents and by the plasticizer in Plastisol coatings.

For sealed glazing units, frame sealants, glazing sealants and gaskets, see HAPM Component Life Manual p. 4.11-4.13.

Key failure modes

UV degradation, weathering/surface erosion, chafing/pull out around fixings (wind/thermal movement), fracture/indentation due to imposed loading/impact.

Key durability issues

UV protection, surface protection, material thickness, adequacy of fixings/supports, exposure conditions.

Balustrades - infill panels

BPG | **5 - Stairs & Balustrades**

External	Description	Maintenance
	Softwood (external use)	
25	Permeable softwoods (eg Scots Pine (Redwood)) pressure impregnated with CCA to BWPA schedule P2 or with creosote to schedule T4. Non permeable softwoods (eg Douglas Fir, Hemlock, Spruce) pressure impregnated with CCA to BWPA schedule P4 or with creosote to schedule T5. Any joints fully coated in 'WBP' quality adhesive or 'D4' adhesive to BS EN 204.	Redecorate: stain every 3 years or paint every 5 years. Renew any creosote coating every 3 years.
25	Permeable softwoods double vacuum impregnated with organic solvent to BWPA schedule V1 or V2. Non permeable softwoods, or mixed species, double vacuum impregnated with organic solvent to BWPA schedule V3. Any joints fully coated in 'WBP' quality adhesive or 'D4' adhesive to BS EN 204.	Redecorate: stain every 3 years or paint every 5 years. Renew any creosote coating every 3 years.
20	Heartwood only of untreated softwood of a species designated as 'SW' or 'SWC' (suitable) for external use without preservation in Table NA1 of BS EN 942 (eg Western Red Cedar and American Douglas Fir). Any joints fully coated in 'WBP' quality adhesive or 'D4' adhesive to BS EN 204.	Redecorate: stain every 3 years or paint every 5 years. Renew any creosote coating every 3 years.
10	Non permeable softwoods, or mixed species, double vacuum impregnated with organic solvent to BWPA schedule V1 or V2. Or pressure impregnated with CCA or BWPA schedule P2 or creosote to BWPA schedule T4. Any joints fully coated in 'MR' quality adhesive or 'D3' adhesive to BS EN 204.	Redecorate: stain every 3 years or paint every 5 years. Renew any creosote coating every 3 years.
10	Permeable softwoods dipped/immersed in organic solvent for a minimum of 3 minutes. Any joints fully coated in 'MR' quality adhesive or 'D3' adhesive to BS EN 204.	Redecorate: stain every 3 years or paint every 5 years. Renew any creosote coating every 3 years.
5	Non permeable softwoods, or mixed species, dipped/immersed in organic solvent for a minimum of 3 minutes. Any joints fully coated in 'MR' quality adhesive or 'D3' adhesive to BS EN 204.	Redecorate: stain every 3 years or paint every 5 years. Renew any creosote coating every 3 years.
5	Untreated softwoods of a species designated as 'SP' or 'SPC' (only suitable for external use if preservative treated) in Table NA1 of BS EN 942 (eg Hemlock, Scots Pine, European Redwood) or softwood only treated by brushing or untimed immersion. Any joints fully coated in 'MR' quality adhesive or 'D3' adhesive to BS EN 204.	Redecorate: stain every 3 years or paint every 5 years. Renew any creosote coating every 3 years.
U1	Unclassified, ie untreated softwoods of indeterminate species or of a species designated as 'X' (unsuitable) for external use in Table NA1 of BS EN 942 (eg Parana Pine).	Unclassified.

For Adjustment factors, Assumptions, Key failure modes and Key durability issues please see overleaf.

5.4c

Balustrades - infill panels

LOCATIONS - External

Adjustment factors

Stained: -5 years.

Assumptions

All timber specifications will be deemed to mean both heartwood and sapwood unless heartwood explicitly required.

PERMEABLE species means 'easy to treat' and 'moderately easy to treat', NON PERMEABLE species means 'difficult to treat' and 'extremely difficult to treat' as defined in BS EN 350-2.

Organic solvent preservatives to be to BWPA Table 4 Type F/N (ie fungicidal and insecticidal).

References to BWPA treatment schedules means schedules included in the British Wood Preserving Association Manual.

Designed and constructed in accordance with BS 6180.

General timber quality must be defined in the job specification either by reference to BS EN 942 (Classes J2-J50) or by a suitable structural timber specification (to limit knots, fast grown timber, grain slope and the like).

In the majority of cases specifications for structural timbers will only make reference to strength grades (eg spruce-pine-fir); GS, MGS, SC3 and the like) which will inevitably mean mixed species which will include permeable and non permeable types.

Fixings and connectors to be plated or non-ferrous.

Key failure modes

Fungal/insect attack, cracks/splits, open joints, distortion, movement.

Key durability issues

Timber species & permeability, preservative treatment, surface protection, maintenance frequency, adhesive type, exposure conditions.

Balustrades - infill panels

BPG   5 - Stairs & Balustrades

External	Description	Maintenance
	Hardwood (external use)	
30	Permeable hardwoods (eg Opepe, English Elm) pressure impregnated with CCA to BWPA schedule P2 or with creosote to schedule T4. Non permeable hardwoods (eg Idigbo, Luan, Mahogany, Meranti, European Oak, Utile) pressure impregnated with CCA to BWPA schedule P4 or with creosote to schedule T5. Any joints fully coated in 'WBP' quality adhesive or 'D4' adhesive to BS EN 204.	Redecorate: stain every 3 years or paint every 5 years. Renew any creosote coating every 3 years.
30	Permeable hardwoods double vacuum impregnated with organic solvent to BWPA schedule V1 or V2. Non permeable hardwoods, or mixed species, double vacuum impregnated with organic solvent to BWPA schedule V3. Any joints fully coated in 'WBP' quality adhesive or 'D4' adhesive to BS EN 204.	Redecorate: stain every 3 years or paint every 5 years. Renew any creosote coating every 3 years.
25	Heartwood only of untreated hardwood of a species designated as 'SW' or 'SWC' (suitable) for external use without preservation in Table NA2 of BS EN 942 (eg Afromosia, Iroko, Mahogany [but not 'Philippine' mahogany], European Oak, Sapele, Teak, Utile). Any joints fully coated in 'WBP' quality adhesive or 'D4' adhesive to BS EN 204.	Redecorate: stain every 3 years or paint every 5 years. Renew any creosote coating every 3 years.
15	Non permeable hardwoods, or mixed species, double vacuum impregnated with organic solvent to BWPA schedule V1 or V2. Or pressure impregnated with CCA or BWPA schedule P2 or creosote to BWPA schedule T4. Any joints fully coated in 'MR' quality adhesive or 'D3' adhesive to BS EN 204.	Redecorate: stain every 3 years or paint every 5 years. Renew any creosote coating every 3 years.
10	Permeable hardwoods dipped/immersed in organic solvent for a minimum of 3 minutes. Any joints fully coated in 'MR' quality adhesive or 'D3' adhesive to BS EN 204.	Redecorate: stain every 3 years or paint every 5 years. Renew any creosote coating every 3 years.
5	Untreated hardwoods of a species designated as 'SP' or 'SPC' (only suitable for external use if preservative treated) in Table NA2 of BS EN 942 (eg Luan, Meranti, Agba) or hardwood only treated by brushing or untimed immersion. Any joints fully coated in 'MR' quality adhesive or 'D3' adhesive to BS EN 204.	Redecorate: stain every 3 years or paint every 5 years. Renew any creosote coating every 3 years.
U1	Unclassified, ie untreated hardwoods of indeterminate species or of a species designated as 'X' (unsuitable) for external use in Table NA2 of BS EN 942 (eg Ash, Beech, Ramin, Sycamore).	Unclassified.

For Adjustment factors, Assumptions, Key failure modes and Key durability issues please see overleaf.

5.4d

Balustrades - infill panels

Adjustment factors

Stained: -5 years.

Assumptions

All timber specifications will be deemed to mean both heartwood and sapwood unless heartwood explicitly required.

PERMEABLE species means 'easy to treat' and 'moderately easy to treat', NON PERMEABLE species means 'difficult to treat' and 'extremely difficult to treat' as defined in BS EN 350-2.

Organic solvent preservatives to be to BWPA Table 4 Type F/N (ie fungicidal and Insecticidal).

References to BWPA treatment schedules means schedules included in the British Wood Preserving Association Manual.

Designed and constructed in accordance with BS 6180.

General timber quality must be defined in the job specification either by reference to BS EN 942 (Classes J2-J50) or by a suitable structural timber specification (to limit knots, fast grown timber, grain slope and the like).

In the majority of cases specifications for structural timbers will only make reference to strength grades (eg spruce-pine-fir); GS, MGS, SC3 and the like) which will inevitably mean mixed species which will include permeable and non permeable types.

Fixings and connectors to be plated or non-ferrous.

Key failure modes

Fungal/insect attack, cracks/splits, open joints, distortion, movement.

Key durability issues

Timber species & permeability, preservative treatment, surface protection, adhesive type, maintenance frequency, exposure conditions.

5.4d

Balustrades - infill panels ___

LOCATIONS - Internal

BPG	5 - Stairs & Balustrades

Description

Internal	

Softwood (internal use)

35+   Softwood to BS EN 942 Table NA1 designated as suitable for use in internal joinery work.

5   Unspecified timber species or quality. Not to BS EN 942 Table NA1 and/or not suitable for internal joinery.

Hardwood (internal use)

35+   Hardwood to BS EN 942 Table NA2 designated as suitable for use in internal joinery work.

5   Unspecified timber species or quality. Not to BS EN 942 Table NA2 and/or not suitable for internal joinery.

Maintenance

Redecorate: stain every 5 years, paint every 10 years.

Redecorate: stain every 5 years, paint every 10 years.

Redecorate: stain every 5 years, paint every 10 years.

Redecorate: stain every 5 years, paint every 10 years.

Key failure modes

Fungal/insect attack, cracks/splits, open joints, distortion, movement.

Key durability issues

Timber species & permeability, preservative treatment, surface protection, adhesive type, maintenance frequency.

Adjustment factors

None.

Assumptions

Designed and constructed in accordance with BS 6180.

Joinery to be manufactured from timber dried down to the appropriate moisture content for in-service conditions, ie 13-17% for intermittent heating or 8-12% for continuous heating.

Joinery to be transported and stored under cover and well ventilated. Internal joinery must not be installed until the building is watertight.

5.4e

BPG	5 - Stairs & Balustrades		

Internal	External	Description	Maintenance
		Plywood	
35+	25	Marine plywood to BS 1088, bonded with 'WBP' quality adhesive to BS 6566:Part 8.	Redecorate: stain every 3 years, paint every 5 years (external), stain every 5 years, paint every 10 years (internal)
35	15	Plywood to BS EN 636-3 (for use in unprotected external conditions), ie bond quality class 3 to BS EN 314-2 and durability hazard class 3 to BS EN 335-3.	Redecorate: stain every 3 years, paint every 5 years (external), stain every 5 years, paint every 10 years (internal)
30	5	Plywood to BS EN 636-2 (for use in humid conditions), ie bond quality class 2 to BS EN 314-2 and durability hazard class 2 to BS EN 335-3.	Redecorate: stain every 3 years, paint every 5 years (external), stain every 5 years, paint every 10 years (internal)
20	U1	Plywood to BS EN 636-1 (for use in dry conditions with no risk of wetting), ie bond quality class 1 to BS EN 314-2 and durability hazard class 1 to BS EN 335-3.	Redecorate: stain every 3 years, paint every 5 years (external), stain every 5 years, paint every 10 years (internal)
U2	U2	Unclassified, ie plywood not to BS EN 636 or type/designation/adhesive bonding unknown.	Unclassified.

Adjustment factors

Stained plywood for external use: -5 years (except for marine plywood).

Assumptions

Stained or painted with full alkyd paint system (2+1 coats or 1+2 coats).

Notes

BRE Digest 323 provides guidance on selecting wood based panel products.

All plywood panels complying with BS EN 636 are marked with the standard number and the thickness.

US and Canadian plywoods complying with other standards are also available.

Key failure modes

Fungal/insect attack, delamination, cracks/splits, distortion, movement.

Key durability issues

Timber species & permeability, preservative treatment, surface protection, adhesive type, maintenance frequency.

5.4f

Internal Fixtures and Fittings

Internal Fixtures and Fittings

Kitchen furniture

Scope

This section provides data on stainless steel kitchen furniture for non-domestic building types. It includes sink units, tables, benches, worktops, cabinets, racking and shelving. Domestic sinks are covered in the HAPM Component Life Manual and are excluded from this section.

The following component sub-types are included within this section:

	Page
• Stainless steel sink units/tables/benches/worktops	6.1
• Stainless steel cabinets	6.1a
• Stainless steel racking and shelving	6.1b

Standards cited

BS 729: 1971 (1994)	Specification for hot-dip galvanised coatings on iron and steel articles
BS 1449	Steel plate, sheet and strip
Part 2: 1983	Specification for stainless steel and heat-resisting steel plate, sheet and strip
BS EN 10088: 1995	Stainless steels

Other references/information sources

HAPM Component Life Manual page 5.7 (domestic sinks).

Kitchen furniture

LOCATIONS - General

BPG | 6 - Internal Fixtures & Fittings

General	Description		Maintenance
	Stainless steel sink units/tables/benches/worktops		
30	Minimum 1.5mm thick austenitic stainless steel, minimum grade 304 to BS 1449:Part 2/ BS EN 10088 grade 1.4301. Metal legs/supports to be minimum grade 304 stainless steel.		Regular non-abrasive cleaning. Avoid the use of coarse abrasive cleaners and strong acids. Fine scouring powder may be used for stain removal.
25	Minimum 1.2mm thick austenitic stainless steel, minimum grade 304 to BS 1449:Part 2/ BS EN 10088 grade 1.4301. Metal legs/supports to be minimum grade 304 stainless steel.		Regular non-abrasive cleaning. Avoid the use of coarse abrasive cleaners and strong acids. Fine scouring powder may be used for stain removal.
20	Minimum 0.9mm thick austenitic stainless steel, minimum grade 304 to BS 1449:Part 2/ BS EN 10088 grade 1.4301. Metal legs/supports to be minimum grade 400 series (ferritic) stainless steel.		Regular non-abrasive cleaning. Avoid the use of coarse abrasive cleaners and strong acids. Fine scouring powder may be used for stain removal.
15	Minimum 0.9mm thick austenitic stainless steel, minimum grade 304 to BS 1449:Part 2/ BS EN 10088 grade 1.4301. Metal legs/supports to be minimum 610g/m² hot dip galvanized steel to BS 729.		Regular non-abrasive cleaning. Avoid the use of coarse abrasive cleaners and strong acids. Fine scouring powder may be used for stain removal.
5	Austenitic stainless steel, minimum grade 304 to BS 1449:Part 2/ BS EN 10088 grade 1.4301. Thickness less than 0.9mm. Metal legs/supports to be minimum 460g/m² hot dip galvanized steel to BS 729.		Regular non-abrasive cleaning. Avoid the use of coarse abrasive cleaners and strong acids. Fine scouring powder may be used for stain removal.
U1	Unclassified, ie stainless steel of less than BS 1449:Part 2 grade 304/BS EN 10088 grade 1.4301, and/or metal legs/supports less than 460g/m² galvanized steel.		Unclassified.

Adjustment factors

BS 1449:Part 2 grade 316 stainless steel (BS EN 10088 grade 1.4401), minimum 1.5mm thick: +5 years.

Regular exposure to chemical agents such as acids, reducing agents (eg laboratory use): -5 years.

Equipment mounted on castors: -5 years.

Units with a bolt together construction: -5 years.

Assumptions

Metal legs to be provided with castors or adjustable feet.

Fixings to be at least galvanized steel or non-ferrous and compatible with other metals used.

Underside of worktops stiffened at regular intervals with metal box sections or equivalent.

Plastic protective layer to be retained until building works (including tiling) completed.

Manufacturers' maximum recommended loadings not to be exceeded.

Notes

For stainless steel domestic sinks, see HAPM Component Life Manual p.5.7.

Key failure modes

Corrosion, staining, scratching, impact/indentation, overloading.

Key durability issues

Stainless steel grade, type/thickness of surface protection (ie galvanized components), cleaning methods, material thickness, type of construction (eg bolted/welded).

6.1

Kitchen furniture

BPG | 6 - Internal Fixtures & Fittings

General	Description	Maintenance
	Stainless steel cabinets	
30	Minimum 1.2mm thick austenitic stainless steel carcase/shelves, minimum grade 304 to BS 1449:Part 2/ BS EN 10088 grade 1.4301. Double skin doors with waterproof core. Metal legs/supports to be minimum grade 304 stainless steel.	Regular non-abrasive cleaning. Avoid the use of coarse abrasive cleaners and strong acids. Fine scouring powder may be used for stain removal.
25	Minimum 0.9mm thick austenitic stainless steel carcase/shelves, minimum grade 304 to BS 1449:Part 2/ BS EN 10088 grade 1.4301. Double skin doors with waterproof core. Metal legs/supports to be minimum grade 304 stainless steel.	Regular non-abrasive cleaning. Avoid the use of coarse abrasive cleaners and strong acids. Fine scouring powder may be used for stain removal.
20	Minimum 0.9mm thick austenitic stainless steel carcase/shelves, minimum grade 304 to BS 1449:Part 2/ BS EN 10088 grade 1.4301. Single skin doors. Metal legs/supports to be minimum 400 series (ferritic) stainless steel.	Regular non-abrasive cleaning. Avoid the use of coarse abrasive cleaners and strong acids. Fine scouring powder may be used for stain removal.
15	Minimum 0.9mm thick austenitic stainless steel carcase/shelves, minimum grade 304 to BS 1449:Part 2/ BS EN 10088 grade 1.4301. Single skin doors. Metal legs/supports to be minimum 460g/m² hot dip galvanized steel to BS 729.	Regular non-abrasive cleaning. Avoid the use of coarse abrasive cleaners and strong acids. Fine scouring powder may be used for stain removal.
10	Austenitic stainless steel, minimum grade 304 to BS 1449:Part 2/ BS EN 10088 grade 1.4301. Metal legs/supports to be minimum 460g/m² hot dip galvanized steel to BS 729. Thickness less than 0.9mm.	Regular non-abrasive cleaning. Avoid the use of coarse abrasive cleaners and strong acids. Fine scouring powder may be used for stain removal.
U1	Unclassified, ie stainless steel of less than BS 1449:Part 2 grade 304/BS EN 10088 grade 1.4301, and/or metal legs less than 460g/m² galvanized steel.	Unclassified.

Adjustment factors

BS 1449:Part 2 grade 316 stainless steel (BS EN 10088 grade 1.4401), minimum 1.2mm thick: +5 years.

Equipment mounted on castors: -5 years.

Assumptions

Metal legs to be provided with castors or adjustable feet.

Fixings to be at least galvanized steel or non-ferrous and compatible with other metals used.

Stainless steel shelves, drawers and plinths to be minimum grade 304 to BS 1449:Part 2/BSEN 10088 grade 1.4301.

Plastic protective layer to be retained until building works (including tiling) completed.

Manufacturers' maximum recommended loadings not to be exceeded.

Key failure modes

Corrosion, staining, scratching, impact/indentation, overloading.

Key durability issues

Stainless steel grade, type/thickness of surface protection (ie galvanized components), cleaning methods, material thickness, type of construction (eg bolted/welded).

6.1a

Kitchen furniture

LOCATIONS - General

6 - Internal Fixtures & Fittings

General	Description	Maintenance
	Stainless steel racking & shelving	
30	Minimum 1.2mm thick austenitic stainless steel shelves, minimum grade 304 to BS 1449:Part 2/ BS EN 10088 grade 1.4301. Metal legs/supports to be minimum grade 304 stainless steel.	Regular non-abrasive cleaning. Avoid the use of coarse abrasive cleaners and strong acids. Fine scouring powder may be used for stain removal.
25	Minimum 0.9mm thick austenitic stainless steel shelves, minimum grade 304 to BS 1449:Part 2/ BS EN 10088 grade 1.4301. Metal legs/supports to be minimum grade 304 stainless steel.	Regular non-abrasive cleaning. Avoid the use of coarse abrasive cleaners and strong acids. Fine scouring powder may be used for stain removal.
20	Minimum 0.9mm thick austenitic stainless steel shelves, minimum grade 304 to BS 1449:Part 2/ BS EN 10088 grade 1.4301. Metal legs/supports to be minimum 400 series (ferritic) stainless steel.	Regular non-abrasive cleaning. Avoid the use of coarse abrasive cleaners and strong acids. Fine scouring powder may be used for stain removal.
15	Minimum 0.9mm thick austenitic stainless steel shelves, minimum grade 304 to BS 1449:Part 2/ BS EN 10088 grade 1.4301. Metal legs/supports to be minimum 460g/m² hot dip galvanized steel to BS 729.	Regular non-abrasive cleaning. Avoid the use of coarse abrasive cleaners and strong acids. Fine scouring powder may be used for stain removal.
10	Austenitic stainless steel shelves, minimum grade 304 to BS 1449:Part 2/ BS EN 10088 grade 1.4301. Thickness less than 0.9mm. Metal legs/supports to be minimum 460g/m² hot dip galvanized steel to BS 729.	Regular non-abrasive cleaning. Avoid the use of coarse abrasive cleaners and strong acids. Fine scouring powder may be used for stain removal.
U1	Unclassified, ie stainless steel shelves of less than BS 1449:Part 2 grade 304/BS EN 10088 grade 1.4301, and/or metal legs/supports less than 460g/m² galvanized steel.	Unclassified.

Adjustment factors

BS 1449:Part 2 grade 316 stainless steel (BS EN 10088 grade 1.4401), minimum 1.2mm: +5 years.

Equipment mounted on castors: -5 years.

Assumptions

Metal legs to be provided with castors or adjustable feet.

Fixings to be at least galvanized steel or non-ferrous and compatible with other metals used.

Plastic protective layer to be retained until building works (including tiling) completed.

Manufacturers' maximum recommended loadings not to be exceeded.

Key failure modes

Corrosion, staining, scratching, impact/indentation, overloading.

Key durability issues

Stainless steel grade, type/thickness of surface protection (ie galvanized components), cleaning methods, material thickness, type of construction (eg bolted/welded).

6.1b

Alphabetical index of components

Appendices

Appendices

Appendix A

Abbreviation	Full name	Abbreviation	Full name	Abbreviation	Full name
BBA	British Board of Agrement	DD	Draft for Development	HAPM	Housing Association Property Mutual
BPG	Building Performance Group Ltd.	DPC	Damp-proof course		
BRE	Building Research Establishment	DPM	Damp-proof membrane	LHC	London Housing Consortium
BS	British Standard				
BSI	British Standards Institution	EN	Euronorm	PSA	Property Services Agency
BWPDA	British Wood Preserving and Damp-proofing Association	FeRFA	Federation of Specialist Contractors & Materials Suppliers to the Construction Industry	PVC	Polyvinyl chloride
				PVC-u	Unplasticised polyvinyl chloride
CIRIA	Construction Industry Research and Information Association			TRADA	Timber Research and Development Association
CP	Code of Practice	GGF	Glass and Glazing Federation		
CWCT	Centre for Window & Cladding Technology	GRP	Glass fibre-reinforced plastic	UV	Ultra-violet.

Appendix B

Definitions/adjustments for exposure conditions

The following definitions are intended to aid users of the Manual to determine whether any adjustment factors apply to the particular buildings or schemes with which they are concerned.

"Normal environment" - Life assessments are provided on the basis of a normal environment, which is assumed to be inland, with normal urban atmospheric pollution only.

"Polluted/industrial environment" - An environment with airborne sulphur dioxide, acid or alkali pollution, normally from an industrial source. The Ministry of Agriculture, Fisheries and Food publishes a map every five years showing the average atmospheric corrosivity rate for 10 km grid squares of the UK. This should be taken as the basis for the assessment, with no adjustment made for microclimatic differences.

"Marine environment" - Coastal areas subject to salt spray and/or sea water splashes. These may extend up to 3 km from the coast or tidal estuary depending on prevailing wind and topography.

"High risk frost locations" - A site is very liable to frost where the following factors all apply:

i) The average annual frost incidence is in excess of 60 days.
ii) The average annual rainfall is in excess of 1000 mm.
iii) The altitude of the site is in excess of 91 m above sea level.

"With 3rd party assurance" - A product with a certificate indicating that ongoing testing and assessment of the product's suitability and/or adherence to claimed standards is carried out by an independent third party such as the British Board of Agreement, Timber Research and Development Association, etc. Quality certification to BS EN ISO 9000 does not match this definition. BSI kitemarking does entail such independent ongoing testing.

"Damp/humid environments" - Internal environments subject to frequent and/or severe wetting or condensation, eg kitchens, bath/shower rooms, laundries.

B1

Appendix C

BPG | List of components included in HAPM Manual

C1

BPG List of components included in HAPM Manual

BPG **List of components included in HAPM Manual**

Appendix C

3 – Roofing components

4 – Doors, windows and joinery

5 – Mechanical equipment components

Appendix C

5 – Mechanical equipment components

6 – Electrical equipment components

Appendix C

6 – Electrical
equipment
components

7 – External
works and
outbuildings

Printed and bound by CPI Group (UK) Ltd, Croydon, CR0 4YY
01/11/2024
01782610-0020